SpringerBriefs in Applied Sciences and Technology

SpringerBriefs in Computational Intelligence

Series Editor

Janusz Kacprzyk, Systems Research Institute, Polish Academy of Sciences, Warsaw, Poland

SpringerBriefs in Computational Intelligence are a series of slim high-quality publications encompassing the entire spectrum of Computational Intelligence. Featuring compact volumes of 50 to 125 pages (approximately 20,000–45,000 words), Briefs are shorter than a conventional book but longer than a journal article. Thus Briefs serve as timely, concise tools for students, researchers, and professionals.

More information about this subseries at http://www.springer.com/series/10618

Parikshit N. Mahalle · Pooja A. Shelar ·
Gitanjali R. Shinde · Nilanjan Dey

The Underwater World for Digital Data Transmission

 Springer

Parikshit N. Mahalle
Department of Computer Engineering
Smt. Kashibai Navale College
of Engineering
Pune, Maharashtra, India

Pooja A. Shelar
Department of Computer Engineering
Smt. Kashibai Navale College
of Engineering
Pune, Maharashtra, India

Gitanjali R. Shinde
Department of Computer Engineering
Smt. Kashibai Navale College
of Engineering
Pune, Maharashtra, India

Nilanjan Dey
Department of Computer Science
and Engineering
JIS University
Kolkata, West Bengal, India

ISSN 2191-530X   ISSN 2191-5318   (electronic)
SpringerBriefs in Applied Sciences and Technology
ISSN 2625-3704   ISSN 2625-3712   (electronic)
SpringerBriefs in Computational Intelligence
ISBN 978-981-16-1306-7   ISBN 978-981-16-1307-4   (eBook)
https://doi.org/10.1007/978-981-16-1307-4

This Springer imprint is published by the registered company Springer Nature Singapore Pte Ltd.
The registered company address is: 152 Beach Road, #21-01/04 Gateway East, Singapore 189721, Singapore

Preface

> *Any Sufficiently Advanced Technology Is Equivalent to Magic*
>
> Arthur C. Clarke

We can annotate underwater world as a magical world, if commonly used as a wireless digital data transmission medium. Water is the largest resource available to mankind, but technologically, it is the least used resource. The research community can forecast an day when terrestrial wireless or wired networks will be insufficient to fullfill the worlds internet demand. At this point the underwater wireless sensor networks can serve the need of world. Therefore the fifth chapter of book focusses more on the application perspective of underwater wireless networks.

This book to our knowledge is the first to discuss all networking aspects required for transmitting digital data through underwater medium under one single roof. It starts from a basic overview of UWSN up to threat attacks and future of UWSN.

The book is arranged into six chapters: Chapter 1 describes the characteristics, history and each single component of underwater wireless sensor networks like acoustic modems, buoys, AUVs, ROVs, etc. Chapter 1 also clearly delineates the difference between terrestrial and underwater wireless sensor networks. Then, it goes over the classification of UWSN and then the diagrammatic explanations on one-, two-, three- and four-dimensional architectures of UWSN. The first part of the book is enough for beginners to inculcate themselves with basic knowledge of UWSN.

Chapter 2 starts with wired and wireless underwater sensor network, and then, it further extends by describing three possible wireless underwater communication mediums (radio, acoustic and optical). This chapter also makes a note of different challenges faced by wireless communication medium. The key part of the second chapter is the differentiation of underwater communication mediums, which will help underwater application developers for selecting appropriate communication medium for their underwater projects.

Then in Chap. 3 of the book, we go one step deeper by discussing the four-layer protocol stack of UWSN nodes. The mapping of routing strategies to different application scenarios is also one of the key parts of the book. The chapter gives a short overview on the functionality and protocols developed in each layer.

The first three chapters will help the readers to understand "what is UWSN and how data can be transmitted through UWSN". But any type of data transmission, terrestrial or underwater, is incomplete without providing security to it. Therefore, Chap. 4 of the book focuses on threats and attacks in UWSN. The data security in UWSN is the least explored research area, and thus, this part is the main body of the book which highlights nearby 32 different types of attacks on UWSN. The cross-layer approach for effective integration of all UWSN communication functionalities is also described in this part, which is a "feather on hat" like feature of UWSN. The best part of the book is that it also made a note of all attack resistance strategies developed so far in state of the art. Thus, the book reduces the readers' time and energy of studying multiple research papers.

Chapter 5 of the book shows the practicality of UWSN by discussing different range of underwater applications starting from surveillance applications like seaweb to environment monitoring applications like DART. The other prototype projects like Microsoft Natick is also covered in this part of the book.

The concluding chapter of the book has mentioned some open research issues in transmitting digital data through underwater world so as to increase the attention of prominent researchers all over the world. This authoritative resource is covering wide range of topics related to UWSN and thus will prove to be a valuable resource for researchers and practitioners in UWSN.

Pune, India   Parikshit N. Mahalle
Pune, India   Pooja A. Shelar
Pune, India   Gitanjali R. Shinde
Kolkata, India   Nilanjan Dey

Acknowledgements

We would like to thank many people who encouraged and helped us in various ways throughout this book, namely our colleagues, friends and students. Special thanks to our family for their support and care.

We are thankful to honourable Founder President of STES, Prof. M. N. Navale, Founder Secretary of STES; Dr. Mrs. S. M. Navale, Vice President (HR); Mr. Rohit M. Navale, Vice President (Admin); Ms. Rachana M. Navale, our Principal; Dr. A. V. Deshpande, Vice Principal; Dr. K. R. Borole; and Dr. K. N. Honwadkar for their constant encouragement and inexplicable support.

We are also very much thankful to all our department colleagues at SKNCOE, JIS University, and for their continued support, help and keeping us smiling all the time.

Last but not least, our acknowledgements would remain incomplete if we do not thank the team of Springer Nature who supported us throughout the development of this book. It has been a pleasure to work with the SpringerBriefs team, and we extend our special thanks to the entire team involved in the publication of this book.

Parikshit N. Mahalle
Pooja A. Shelar
Gitanjali R. Shinde
Nilanjan Dey

Contents

About the Authors

Dr. Parikshit N. Mahalle is Professor and Head of Department of Computer Engineering at STES's Smt. Kashibai Navale College of Engineering, Pune, India. He has his Ph.D. from Aalborg University, Denmark, and continued as Postdoc Researcher. He has 20+ years of teaching and research experience. He is Member of Board of Studies in Computer Engineering SPPU and various Universities. He has 7 patents, 130+ research publications (citations-1412, H index-16) and authored/edited 20+ books with Springer, CRC Press, Cambridge University Press, etc. He is Editor-in-Chief for IGI Global—*International Journal of Rough Sets and Data Analysis*, Associate Editor for IGI Global—*International Journal of Synthetic Emotions, Inderscience International Journal of Grid and Utility Computing*, and Member of Editorial Review Board for IGI Global—*International Journal of Ambient Computing and Intelligence*. His research interests are algorithms, Internet of things, identity management, and security. He has delivered 100 lectures at national and international levels.

Pooja A. Shelar is pursuing her Ph.D. from Savitribai Phule Pune University. She has her M.E. (Computer Engineering) and B.E. (Computer Engineering) degrees from SPPU. She has worked on projects related to secured sharing of digital data and images in wireless sensor networks and published 1 book chapter and 4 research papers in referred & indexed journals, and conferences at international and national levels. Her research interests are wireless sensor networks, information security, and cryptographic algorithms.

Dr. Gitanjali R. Shinde has 11 years of experience, currently working as SPPU approved Assistant Professor in the Department of Computer Engineering, Smt. Kashibai Navale College of Engineering, Pune. She has her Ph.D. in Wireless Communication from CMI, Aalborg University, Copenhagen, Denmark. She has her M.E. (Computer Engineering) and B.E. (Computer Engineering) degrees from the University of Pune and received research funding from SPPU. She has presented research article in World Wireless Research Forum (WWRF) meeting, Beijing, China, published 40+ papers in national and international conferences and journals, and authored 5+ books with Springer, CRC press.

Nilanjan Dey is Associate Professor in the Department of Computer Science and Engineering, JIS University, Kolkata, India. He is a visiting fellow of the University of Reading, UK. He held an honorary position of Visiting Scientist at Global Biomedical Technologies Inc., CA, USA (2012–2015). He has his Ph.D. from Jadavpur University. He has authored/edited 80+ books with Elsevier, Wiley, CRC Press, and Springer and published 300+ papers. He is Editor-in-Chief of the International Journal of Ambient Computing and Intelligence (IGI Global), Associated Editor of *IEEE Access and International Journal of Information Technology* (Springer), Series Co-editor of *Springer Tracts in Nature-Inspired Computing* (Springer) and *Advances in Ubiquitous Sensing Applications for Healthcare* (Elsevier), and Series Editor of *Computational Intelligence in Engineering Problem Solving* and *Intelligent Signal Processing and Data Analysis* (CRC). His research interests are medical imaging, machine learning, computer-aided diagnosis, data mining, etc. He is the Indian Ambassador of the International Federation for Information Processing—Young ICT Group—and Senior Member of IEEE.

Chapter 1
Introduction to Underwater Wireless Sensor Networks

1.1 Basics of Underwater Wireless Sensor Networks

"Sensor networks" can be said as a significant part of technology that creates magic of sensing and wirelessly sending information on land and in water. Sensor networks built on land are popularly named as terrestrial sensor networks which are used in many applications from IoT, VANET to smart cities. The network below water is known as "underwater sensor network" and is used for different range of underwater applications such as oceanographic study, disaster prevention system, water quality monitoring systems, surveillance reconnaissance and many more explained in chapter named Applications of UWSN.

The three-fourths of earth surface is covered by water, and hence, water is one of the largest natural resources available for mankind. This fact and increasing worldwide population suggest that it's time to develop a technological underwater world without harming the aquatic species for increasing human needs like shortage of land, requirement of more data centres, need of oil, minerals, metals and so on. In this recent year, UWSN has become one of the emerging technologies and thus it needs to understand the basics of UWSN before going into its detail study.

1.1.1 Underwater Wireless Sensor Network (UWSN)

Definition:
UWSN can be defined as an underwater data network of self-governed sensors and acoustic modems which are capable of sensing and sending digital data in underwater environment, by using sound waves as a basic medium of communication.

UWSN is a sensor network which can be built at the bottom of oceans, lakes or rivers. As shown in Fig. 1.1, the network consists of a gateway buoy at surface, a network of sensor nodes deployed at seafloor and an onshore station connected to

© The Author(s), under exclusive license to Springer Nature Singapore Pte Ltd. 2021
P. N. Mahalle et al., *The Underwater World for Digital Data Transmission*,
SpringerBriefs in Computational Intelligence,
https://doi.org/10.1007/978-981-16-1307-4_1

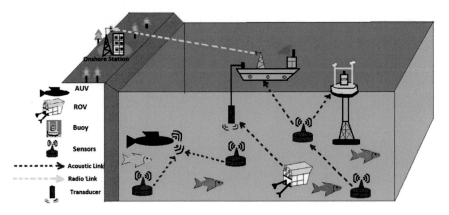

Fig. 1.1 Diagrammatic representation for UWSN

sink node and satellite. The bottom to upflow of data in UWSN starts from the sensor deployed at seafloor to the sink node at upper surface and lastly to the onshore station. The main components of UWSN are acoustic modems, buoy, AUVs, ROVs and onshore station. Optical sensors or some application-specific sensors are also be used in construction of UWSN. Each of this component is detailedly explained in this chapter.

1.1.2   History of UWSN

The very first origin of underwater communication was between 384 and 322 B.C. when Aristotle stated that sound waves can travel in water as well as in air. Then after 2000 years later, i.e. in 1490, a scientist named Leonard da Vinci made an observation that humans can hear the ship over a long distance by just placing one head of the long tube in water and the other end towards the human ears.

Then after 200 years later, i.e. in 1620s, the foundation of acoustic was given by Marin Mersenne who published his work on the speed of sound in air. And then the first mathematical theory of how sound travels in air was given by Issac Newton in 1687. Although the theory was proven for transmission of sound in air, the same basic mathematical theory is also applicable for transmission of sound in water.

In 1743, Abbé J. A. Nollet was the first time to confirm that "Sound Travels in Water". He conducted series of experiments by keeping his head underwater and then listing to the sound of pistol shot, bell, alarm clock and whistle. Then from his observations, he noted that sound can easily be heard by underwater observer.

Then after a long time, in 1826 a physicist and mathematician named Jean-Daniel Colladon and Charles-Francois Strum measured "Speed of Sound in Water". Scientist used Leonardo da Vinci theory of placing long tube in underwater to measure the speed of sound generated by a bell. The experiment was conducted in Lake Geneva with two boats acting as sender and receiver. The first boat was responsible to ring the underwater bell simultaneously with the ignition of gunpowder. The sound of bell and the flash from gunpowder were observed 16.09 km away on the second boat. Speed calculation was done using time between the flash and sound reaching the second boat. They determined the speed of sound in fresh water to be 1435 m/s.

After these inventions, scientist started thinking about the practical applications of underwater sound/acoustic. Thus, in 1838 a scientist named Charles Bonnycastle came up with the very first underwater application which was determining the depth of oceans by using echoes. The water depth was calculated using the time taken by an acoustic signal to reach the bottom and the echo to return back to the ship.

Now around 1854, there was an era of telegraph messaging and there were 20,000 miles of telegraph cables crisscrossing the United Nations. Then an American merchant named Cyrus West Field developed an idea to lay down a well-insulated telegraph cable across the floor of Atlantic Ocean. This invention was the first step towards sending data or messages in underwater environment for the purpose of reducing the network of terrestrial telegraph cables.

In the year of 1857–1858, one more underwater application for voice communication among submarines was invented. The U.S. Navy developed an acoustic system called "underwater telephones" which was able to send and receive human voice using sound waves. This invention had set a benchmark for the development of various different underwater applications.

In the twentieth century, the underwater acoustic got a different direction due to the mathematical theory of sound wave which was given by Lord Rayleigh. The third underwater application was developed in mid-April of 1912, by a company named Submarine Signal. They developed an echo ranging system for submarine communication. This system used an echo ranging device also known as hydrophones to send the dots and dashes of Morse code using acoustic signals. The device is a high-powered underwater loudspeaker which can produce and detect sound. Echo ranging device was developed by Reginald A. Fessenden so in the later stage it was called "Fessenden Oscillator".

Now the year of 1913 was an historic year as it witnessed the underwater data transmission. Because in Boston harbour, Fessenden used new oscillator and was successful in sending messages over several miles between two tugboats. As 1912 was a year when passenger ship Titanic collides with an iceberg, that incident encouraged scientist to think over underwater discoveries. In relation to this incident, a year later Fessenden was able to detect a 130-foot-high and 450-foot-long iceberg more than two miles away. He was also able to detect the ocean depth of 31 fathoms that is 186 ft. In spite of such exceptional results, the submarine signal company did not commercialize the echo ranging system. After ten years later, the company produced a low-frequency echo sounder based on Fessenden oscillator and then renamed it as "fathometer" because depth was measured in fathoms. The underwater telegraph system using Fessenden oscillator had become the most popular system among all submarines communication in 1930s.

Development in underwater acoustic took a fast speed after the increased use of submarines in World War I. From 1914 to 1918 was the period when underwater applications got deeply associated with military applications. The different aspects like security of underwater data transmission, surveillance of water territory and detection of target submarine came into picture during World War I. The submarine detection in World War I was done by using a two-earphone device which was worn by the sonar operator who was able to determine the direction of submarine sound coming from the propellers or engine.

The hydrophone with a piezoelectric effect was used to build an echo ranging system which was discovered by Paul-Jacques and Pierre Curie in 1880. After then a quartz crystal was used to generate sound by Langevin. Then in 1918, it was proven that a sound wave can travel a distance of 1500 m. The commercialization of echo ranging system was done after World War I.

Many basic concepts on underwater sound propagation where studied and cleared by group of scientists during the period between World War I and World War II. In this period, a German scientist brought a new theory on the refraction and bending of sound waves in sea water. The Willebrord Snell in 1919 stated that sound waves get refracted due to any change in temperature, salinity and water pressure. Therefore, the sound wave refraction in underwater medium was the first challenge in underwater data transmission found by a Dutch scientist. He also claimed that ocean currents and climatic changes would affect the way how sound travels in water.

In 1922, the echo sounder was practically used in best route selection for underwater telegraph cables between France, Marseilles and Philippeville. In this same period, scientists were successful in finding oil mines present beneath the seafloor using underwater acoustics. The seismic recorders which can receive sound signal on seafloor were developed by Ewing, Vince and Joe Worzel in 1934.

In 1937, the performance of echo ranging systems deteriorated as it was failing to detect submarines in shadow zone created due to sound wave refraction. Then came a new instrument into picture which was named as bathythermograph which showed that speed of sound in warm water surface that's the upper part of ocean is greater as compare to below colder water surface. This measure gave a boom to research on speed of sound in different water depth.

In World War II, there came a SONAR which is acronym of Sound for Navigation and Ranging. The SONAR technique uses sound wave to communicate or detect objects under the surface of water. Then in 1950s, the U.S. Navy developed SOFAR (Sound Fixing and Ranging) channel for detection of submarines at a distance of 100 miles. Again, in mid-1950s, U.S. Navy developed a SOSUS (**SO**und **SU**rveillance **S**ystem) which had 40 hydrophones connected in a 1000-foot-long horizontal line at a depth of 1440 ft. All the hydrophones were connected by underwater cables which can expanded maximum 241 km long to an onshore station called as "Naval Facilities". After getting success in first prototype, the U.S. Navy decided to build such arrays of hydrophone along the entire US East and West Coast. SOSUS can be considered as first step towards formation of underwater sensor network (UWSN).

The research work on underwater acoustic was going on and in 2000, the submarines were able to send and receive e-mails over a long distance that to without changing their locations or raising the antenna above water surface. This was a revolutionary change in the field of underwater data transmission. The answer to "How data is exchanged in underwater environment?" is explained in the subsequent chapters of the book.

1.1.3   Difference Between TWSN and UWSN

All the possible things in terrestrial wireless sensor network (TWSN) become impossible in underwater wireless sensor network (UWSN). For example, the radio waves are used as communication medium in TWSN but in UWSN radio waves cannot penetrate into water. The water as a communication channel poses many challenges in development and functioning of UWSN. Therefore, the comparative study shown in Table 1.1 will help in better understanding of underwater networks.

1.2   UWSN: A Need of Today's World

The UWSN is gradually becoming a world-changing technology and indirectly as a necessity. The term "world changing" is given because its UWSN which is making it possible to develop almost all things below water which mankind has develop on earth surface. The plan of developing an underwater city by a Japanese company can be one of the examples of a world-changing UWSN. The Japanese company has claimed to build underwater five star hotels, shopping malls and residential houses in the coming five years. If this is the future, then it is a need to focus more and explore the field of underwater data transmission.

The world's population is increasing every day, but the land space available is correspondingly less. Therefore, there is a need to save land surface and utilize the world's largest natural resource, i.e. surface below water. For this reason, Microsoft has developed a prototype of underwater data centre. Imagine if all data centres are placed below water, then it will indirectly reduce the huge land requirement.

The increasing wired underwater network makes it another necessity to develop a wireless underwater network. The journey of underwater cable started from 1850s, when the very first underwater cable for telegram was laid. And now there are 300 undersea cables stretching 550,000 miles. The bottom of oceans is filled with the world connecting Internet cables. The deployment cost of one cable is millions of dollars, and it takes months to lay down one cable. Therefore, it is a need to breakdown the increasing underwater wired cable network and create a low cost and low maintenance solution using underwater wireless network for data transmission. The breakdown of wires is also important for conducting short experiments in oceans. For example, if a scientist wants to test a prototype module

Table 1.1 Comparative study of TWSN and UWSN

Parameters	Description	TWSN	UWSN	Reason
Communication medium	It is means of transmission along which information is send and receive	Radio waves	Acoustic wave, optical wave	High-frequency radio waves cannot penetrate in salty water as they have tendency of attenuation
Storage capacity	Its caching capacity of sensor node to cache sense data	Less	More	As sensor replacement is difficult in UWSN
Bandwidth	Maximum frequency at which data can be transferred over the communication medium	High (in MHz)	Low (in KHz)	Bandwidth is distance dependent for UWSN. Shorter the distance greater is the bandwidth vice versa But in TWSN bandwidth is fixed even with varying distance
Frequency	Its number of cycles done per unit time. One cycle is equal to 1 Hz. Each cycle carries bits of information	High (GHZ or MHz)	Low (Hz or KHz)	High-frequency waves are quickly absorbed by water so only low-frequency waves can travel into water
Propagation speed	The speed at which a signal can propagate is said as propagation speed	The speed of radio wave is 300,000 km/s	The speed of acoustic wave is 15,000 m/s in water	Speed of acoustic wave in water is faster as compared to air. The speed also depends on the temperature of water. Higher the temperature of water greater is the propagation speed of acoustic wave
Propagation delay	A small amount of finite time delay taken by data to reach its destination is known as propagation delay	10 μs	50–1000 ms	In UWSN, the communication mediums face challenges like multipath effect and noise which leads to larger propagation delay

(continued)

Table 1.1 (continued)

Parameters	Description	TWSN	UWSN	Reason
Signal transmission rate	The rate at which the information is transferred between sender and receiver is said as signal transmission rate	12,000–9600 bps	10 bps	Doppler spread and path loss are the main reason for very low transmission rate in UWSN
Topology	Topology is a pre-decided structure of paths and sensor positions along which the data will travel	Star, mesh, bus, tree, ring, circular, grid, etc.	Square and triangle topology [12]	Node mobility, propagation delay and energy consumption make the topology difference between TWSN and UWSN
Deployment	Deployment deals with total number of sensors to be used and technique used for deployment	Dense	Sparse	In UWSN, there is a limitation over total number of sensors used. So underwater networks are constructed with few sensors which are sparsely deployed
Node mobility	Depending on network type, the node can be mobile or fixed	Static nodes	Dynamic nodes	Underwater sensors move with water currents
Position information	The sensors use position information for sending data across networks	GPS is used	GPS do not work in underwater networks	The Global Positioning System uses high frequency radio waves which cannot penetrate into water
Energy consumption	The amount of battery or cell used by sensors in sensing, processing and sampling data	Less	More	The underwater sensors require huge amount of energy for data transmission and also battery is not replaceable
Cost	It includes total cost from price of each network components to their maintenance and deployment cost	Cheap	Costly	The components of UWSN are quite costly like AUVs and ROVs also the maintenances charges are more as compared to TWSN

of water quality monitoring system, then it is good to go with faster and low-cost solution of UWSN. The various underwater applications which have become a necessity of mankind like underwater oil field discovery, study of marine life, tsunami detection systems and many more underwater applications require UWSN for their smooth functioning. Therefore, we can summarize that UWSN is not only a topic of research but have become a need of today's world.

1.3 Components of UWSN

All the elements of network structure which altogether help in the functioning of whole network with an aim of successfully sending and receiving data from and to its destination point are known as components of network. As in terrestrial network, servers, routers, network interface card (NIC), client machine, hub, switch and other software and hardware resources are used as network components. In same way for underwater networks, acoustic modem, optical sensors, buoys, AUVs, ROVs and transduces are said as underwater network components. Hence, for better understanding of underwater networks, each component used in UWSN is described in detail.

1.3.1 Acoustic Modem

Acoustic modem is device or part of sensor that can measure physical quantity from environment and then converts it into sound signal. These signals are then used to transmit sensed data at the receiver end. The transmitting range of acoustic sensors is up to few 100 km but is limited on bandwidth. Hence, data transmission done using acoustic modem is said reliable but at low data rate as bandwidth of acoustic wave in underwater environment is minimal. Therefore, acoustic sensors are used in applications which mainly requires large volume of data to be transmitted over long distance. Acoustic wave is the most commonly used communication medium in applications using UWSN, and hence, a number of acoustic sensors deployed are in large amount. Therefore, sometimes UWSN is also said as "underwater acoustic sensor network (UASN)". The sensors used in terrestrial network differ from underwater sensors in many ways, like in power, cost, memory, spatial correlation and development technology used.

Power: The deployed underwater sensor nodes require coordination among their operation, which is achieved by exchanging real-time location and movement information and by sending sensed data to onshore station. Hence, the power consumed by underwater acoustic sensors is more as compared to terrestrial sensors.

Cost: The underwater acoustic sensors are sparsely deployed due to the high cost of acoustic sensors. The modulation technique used in acoustic modem is quadrature amplitude modulation (QAM) and phase shift keying (PSK), which increases the development cost of sensors and thus indirectly increasing overall costing of UWSN.

Memory: The underwater sensors require memory to cache some sensed data, whereas terrestrial sensors do not require large storage area.

Spatial correlation: The sensed data from terrestrial sensors is correlated, but it's unlikely with underwater sensors as the underwater sensors are sparsely deployed.

The acoustic sensor can be developed using three types of modulation techniques non-coherent, coherent or fully coherent. The frequency shift keying (FSK) is a non-coherent modulation technique which is characterized by power-efficiency and low-bandwidth hence are not useful for multi-user networks which require high data rate. Therefore, acoustic modem with coherent modulation technique came into picture which were able to transmit data over a long range. The acoustic modem with fully coherent modulation technique is the recent advancements used in acoustic sensor design. The underwater sensors include sensors to measure the characteristics of water like temperature, oxygen, salinity, pH, hydrogen, turbidity, dissolved methane gas. Sensors are used to measure forces and moments and the underwater quantum sensors to measure light radiation. There are many more underwater sensors developed for various underwater applications.

1.3.2 Optical Sensors

Wireless data transmission in underwater environment is of great importance to many underwater applications like water monitoring systems, tsunami detection, military, oceanography study, oil and gas industry and many more. The common base of all this application is underwater data transmission with high data rate and bandwidth. Now, the most commonly used communication medium for underwater data transmission is acoustic wave, but it is limited by bandwidth. Therefore, requirement of emerging underwater applications and limitation of acoustic link lead to proliferation of underwater optical communication and indirectly use of optical sensors. The working range of underwater optical sensors is 100–200 m for turbid water and up to 300 m in clean water. And the communication bandwidth for optical sensors can be in Mbps or Gbps depending on distance travelled. Optical sensors are battery saver as the energy used for data transmission by optical sensors is less as compared to acoustic modems. The other side of coin which make optical sensors as less used component is the extreme challenges pose on optical link due to the scattering and absorption phenomena of optical wave and also the limitation of distance travelled by optical wave.

The optical sensors use LED-based or laser-based technology for emitting light beam. The LED-based optical sensors give insufficient bandwidth and thus indirectly provide low data rate and low transmission distance. Thus, laser-based optical sensors came into picture which provide large bandwidth and high-speed data rate. Then the optical sensors with data rate of 1 Gbps for distance of 2 m [1] were developed by using green laser with 532 nm wavelength. After period of time, optical sensors with blue laser having 1.4 Gbps of bandwidth over a distance of 4.8 m [2] were developed. Then the optical sensors developed by using quadrature amplitude modulation orthogonal frequency division multiplexing (QAM-OFDM) technique gave a data rate of 4.8 Gbps [3]. The highest data rate of 12.4 Gbps [4] was achieved by using optical sensors with blue laser diode.

One of the examples for optical sensor is "BlueComm 200" [5], which is wireless optical communication system developed to transmit subsea data, stream video at very high speeds. It uses array of LEDs that are rapidly modulated to transmit data. The data transmission range is greater then 150 m having bandwidth 10 Mbps. The BlueComm can also achieve 500 Mbps of bandwidth if operated using blue light region of spectrum.

1.3.3 Autonomous Underwater Vehicle (AUV)

The AUV is an underwater robot which is controlled and piloted by remote computer. The name autonomous because robot is independent of humans and is also not attached to any physical location from which it is launch. AUVs are also sometime known as unmanned underwater vehicle (UUV). The vehicles are free to move and are powered by solar or researchable batteries, and they are also equipped with side-scan sonar, data module, profilers, cameras, sensors, antennas and many more features. However, some large-size vehicles are powered by aluminium-based semi-fuel cells which require more maintenance as compared to battery-operated AUV.

In looking into history page of AUVs, we get to know that a scientist named Robert Whitehead had developed a missile named Automobile Fish Torpedo in 1871. Torpedoes was named after the electric ray fish (also known torpedo) which has an ability to emit electric shocks to its prey. This invention was said as origin of AUV because if we ignore the emission of electric charge then these torpedoes can be called as first AUV. Till date most of the AUVs are still manufactured in traditional torpedo shape because it is best design with a perfect balance of size, usable volume, ease of handling and hydrodynamic efficiency. And the AUV sizes are application depended and it range from portable lightweight AUV to a 10-meter-long AUV. For example, the underwater military application uses a long AUV to endure efficiencies of operating AUV, whereas research institutes and universities use the portable less expensive AUV for making prototype of any underwater application.

AUV Applications:
The autonomous underwater vehicles are used by research community from a decade and is flexible component of UWSN which can be used in many applications like mapping physical structure of ocean, finding submerged cities, for detecting submarine volcanoes. The oil and gas company also use AUV to get a detailed map of seafloor before starting the construction of infrastructure and

deployment of pipelines. The underwater vehicle plays a key role in defence applications for mine detections, hydrography, surveillance and reconnaissance. In defence applications, AUVs are launched and operated from submarines and not from the surface ships. AUV is also used for rapid environmental risk assessment. For example, AUVs can be loaded with many different sensors and thus any change in water chemical molecule can be easily detected using an AUV.

AUV Functioning:
The autonomous vehicles are developed to work on several miles below water surface. The basic idea of how AUV works can be described in few step procedures.

Step 1: The first stage is releasing of AUV on water surface and then making a connection between onboard operator and AUV by sending a radio signal from ship to a transducer on AUV. The connection done is used to set AUV for a preprogrammed trajectory. AUVs are also connected to satellites by using radio signals for the purpose of AUV tracking.

Step 2: After completing connections, AUV starts its journey towards the bottom where it is made to continuously sends its location information to the ship on surface. AUV uses underwater acoustic positioning system during their navigation in seabed.

Step 3: The sensors present in AUV are used to sample ocean and collect superior quality data. All the collected data is stored in data module placed in starting portion of AUV. For example, the side sonar placed in the end part of vehicles is responsible for continuously scanning the path travelled by AUV in search of an object such as an Internet cable placed underwater or submerged cities.

Step 4: The AUV also comprises of different components like a sub-bottom profiler which are used for identifying or characterising sediments or rock under the seafloor whereas the echo-sounder is use for measurement of underwater physical and biological components. The cameras are used for taking pictures of the seafloor or to find the submerged objects in ocean.

Step 5: After travelling for hours in deep oceans, the AUVs ascent back to water surface where the operator recovers the AUV back on ship. And at this point, batteries and data module in vehicles are replaced for reuse of AUV.

Step 6: The last step where the scientist connects the removed data module to a computer for downloading the stored sensed data.

1.3.4 Remotely Operated Underwater Vehicle (ROUV/ROV)

ROUV or most commonly known as ROV is an underwater robot used for sea life learning. Like AUV, the ROV is also an unmanned robot but in addition its connected to an operator on ship using series of cables. The ROVs navigate remotely as per the command and control signals send by the operator through the attached cable. ROVs are mainly equipped with HD video cameras, lightning system, sonar system, articulating arms, water sampler, manipulator, static camera and sensors to measure water clarity, temperature and density. The data sent by ROVs can be in form of a videos, images or digital data which is further useful in many different applications. The ROV is a man-operated robot, whereas AUV is a programmed vehicle independent of operator interaction.

The design of ROVs varies greatly as they range from small size to the size larger than an AUV. Depending on application usage, the ROVs can be divided into four types, i.e. work class ROVs and micro or mini ROVs are used in shallow depths for seafloor exploration, whereas for moderate to deep depth, a light work class ROVs are used. And to explore small lakes, rivers or coastal water, obser-vational class ROVs are used.

ROV Applications:
ROV can go up to 35,000 ft. deepest, whereas a scuba diver can go upon 165 ft. Therefore, ROVs are used in development, inspection, maintenance and repair of offshore oil and gas industry and are replacement for human divers. ROVs are also used for recovering objects from seabed. For example, in 1966 Palomares B-52 crash incident held in Spain in which out of the four nuclear bomb one was lost in Mediterranean Sea. This lost bomb was found after large survey and was then recovered using ROV. The articulating arms of ROVs are used to bring any object or sample from seabed to the operator on ship. ROVs are widely used robot from underwater studies to different industrial applications. They are pioneers in field of archaeological survey, underwater exploration, military, search and rescue, aqua-culture domain.

ROV Functioning:
ROVs are deployed from special ships where the cable attached from ship to ROV acts as an umbilical cord. The ROVs are then maid to perform the specific task by sending them the command signal. These specific tasks can be like inspection of pipeline, finding wrecks or cleaning the mines in case of military applications. The ROVs are brought back onto ships using the pulley; there are generally no time limits for ROV to stay underwater but the maximum time limit is related to battery life of ROVs.

1.3.5   Buoys

The buoys in UWSN act like a bridge between anchored underwater sensors and vehicles at seafloor and the onshore station and satellites. The data collected from sensors crosses the bridge, i.e. surface buoy to reach its destination which is gen-erally onshore station or satellites. Buoys can be found of many different types with each serving a different purpose. For example, mooring buoys are used for mea-suring water properties like salinity, temperature and velocity, whereas the sono-buoys are used for submarine detection and localization. All types of buoys can be either floating or stationary and deployed using aircraft or ship. The buoys can operate in depth of 10–9000 m depending on the need of application.

As the world is transiting in era where information is readily available on finger tips but same is not true with oceanographic data. For digitization of oceans, we need a strong mediator "buoy". In developing UWSN, its important component that is buoy should be able to transmit data in real time as it is collected and then should make data instantly available via Internet or phone. The National Oceanic and Atmospheric Administration (NOAA) [6], an organization in USA, is responsible for deploying and collecting data from each floating buoy. NOAA is managing and maintaining more than 1300 buoys placed in atlantics and pacific oceans. Buoy is constructed using video camera on the top, a communication antenna, navigation light, radar, batteries, solar panel, satellite tracker, cellular satellite WiFi 3G and 4G, float assemble, generator and many sensors and payloads for measuring water and air parameters like wave height, wind direction, water temperature and so on.

Buoys Applications:
Buoy sends real-time data on oceanographic conditions to onshore stations which is beneficial to scientists, public health officials, fishermen, meteorologist and many more. There is a mobile application named "NOAA Smart Buoy" which is freely available an Google Play Store and App store. The mobile app is mainly useful for US fishermen and marine scientists to known the weather conditions in Chesapeake Bay like wind speed, wave height and water and air temperature. Buoys are also used to charge the batteries and to send command to underwater sensors and ROVs. The maritime security is achieved using buoys as they used cameras and radar for surveilling its surrounded water territory. The DART system uses buoys for activating emergency alerts in tsunami, thus making buoy helpful tool in finding and preventing natural disasters. Some buoys are used to convert wave energy into electrical power using a driver generator. This generator continuously charges the energy power system which is used to supply electrical power for power hungry underwater applications.

Buoys Functioning:
The buoys are deployed using a ship or an aircraft. During positioning, the upper yellow colour part of buoy is light weight so becomes floating buoy surface and other parts, i.e. spare tube and heap plate are stationery inside water. After deployment, the buoy is connected to satellite and an onshore station. According to remote station instructions, buoys send real-time oceanographic data and recharges sensors, AUVs and ROV.

1.3.6 Transducers

The sensors are device that sense the physical quantity, whereas the transducers are responsible to convert the one form of signal (non-electrical signal) into another form (usually electrical). Transducers are a part of sensor which convert the sensed data into standard electrical signal. These electrical signals containing data are further used for data communication among sensor and the destination node.

Transducers are also called as sensors when the output of transducer is converted to a readable format. For example, transducer is microphone, light-emitting diode (LED) (converts light energy into electrical), antenna (converts electricity to electromagnetic waves), loudspeaker (converts electrical signal to sound format) and so on. There are many types of transducers such as piezoelectric, pressure, temperature, ultrasonic transducers. The underwater acoustic transducers convert the sound energy into electrical energy. These transducers can contain both transmitter and receiver or only transmitter. They sense reflected sound and use distance and data for communication and navigation. The underwater acoustic transducers can work at unlimited ocean depth, and this device has omnidirectional, horizontal or beam pattern. Many acoustic transducers are arranged together in an array thus forming a system.

The ultrasonic transducers [7] are used by most of the commercial underwater applications which require short range communication. For example, naval requires short-range detection of small objects like torpedoes and mines, real-time ship positioning systems, for fish finding, deep drilling operations, side-scan sonar f or survey operation, wave-height indicator, acoustic imaging and so on. Some of the underwater imaging systems are using micromachined ultrasonic transducer [8] which converts acoustic energy into electrical signal or vice versa. The imaging system requires high-resolution pictures to study or find minute objects. Hence, micromachined ultrasonic transducer provides high bandwidth, high receiving sensitivity and high beam projection for acoustic imaging [9]. Hence, from the literature we can state that transducers are also an import part of underwater applications and underwater sensor network.

1.3.7 On-Shore Station

The on-shore station is the last and the most important component of UWSN. All the components of UWSN like ocean bottom underwater sensor nodes, AUVs, ROVs and buoy generate and send real-time data to a base station/on-shore station. The on-shore station is then responsible to convert all the data into a useful information. The desktop machines used at on-shore station must have high processing capabilities and memory capacity. The researchers are also working on scheduling algorithms [10] for sensor nodes to decide when and how much information is to transmit via acoustic link so the value of data reaching the terrestrial on-shore station is maximized. One of the examples for on-shore station is the Dubai Municipality Corporation [11], which have desktop computers, i.e. base station with Windows operating system. This station collects large amount of real-time data like water temperature, pH, salinity, dissolve oxygen, quantity of chlorophyll. This collected data is then used by Dubai municipality for the purpose of real-time water monitoring.

1.4 Architectural Models in UWSN

Network architecture is a planned organization of the network components in order to perform the application-specific tasks. It's an blue print which showcase the functionalities of network communication framework to design and implement an underwater project. The underwater network architectures cannot be drawn using the traditional terrestrial network architectures because as seen in Table 1.1, UWSN differs from TWSN in many aspects. For example, from architectural point of view the TWSN sensors can be deployed at long distance but in UWSN sensors are deployed over shorter distance because acoustic waves cannot propagate at long distance like electromagnetic waves. Also, in UWSN, the sensor nodes are mobile as they move with ocean currents, leading to change in network topology but the TWSN nodes are static. Most of UWSN requires multiple hops to reach the sink node as the communication range is shorter. But the multi-hop architecture design for UWSN leads to a dilemma that multi-hop architectures will achieve reliable data transfer and at same time increases the energy consumption of sensor nodes. Hence, researchers are working on developing full-fledge solution for UWSN architectures that will deal with issues like node mobility, energy consumption, reliable data transfer and cost efficiency.

The unique characteristics of communication mediums used in UWSN also pose a necessity of designing a different and application-specific network architectures for underwater senor network. As the underwater communication mediums (acoustic, optical and radio waves) have to face various challenges due to underwater environment such as limited bandwidth, propagation delay, absorption, multipath effect and many more. Note that all the communication mediums and their challenges are detailed explained in the next chapter of book. Hence, new innovative research on the architecture and topology control of UWSN will have a great impact on development and performance of various underwater applications.

Classification of UWSN Architecture

The underwater sensor network architecture is classified based on two criteria node mobility and spatial coverage. The node mobility of sensor node means stationery, mobile and hybrid and the other is spatial coverage of sensor node such as two-dimensional or three-dimensional.

Static UWSN: In static underwater sensor architecture, the deployed sensor nodes are fixed to a position; thus, the network topology remains same. The static sensor nodes are deployed by attaching them to surface buoy or anchoring them to ocean floor. The static UWSN is used in monitoring applications which is limited for certain region.

Mobile UWSN: In mobile underwater sensor architecture, sensor nodes move freely with ocean currents and thus topology changes dynamically. There are two types of sensor nodes in mobile UWSN called as unpropelled and propelled nodes.

Unpropelled Sensor nodes:
The mobile underwater sensor architecture considering unpropelled senor node means nodes are not forced to move they float freely and drift with the currents. For example, the drifters, gliders and profiling floats are unpropelled mobile equipment.

Propelled Sensor nodes:
The mobile underwater sensor architecture considering unpropelled senor node means nodes is programmed to move at specific locations or controlled by a remote operator. The AUV and unmanned underwater vehicle are example of propelled sensor nodes and are used for in applications which requires collecting measurements from various layers of ocean.

Hybrid UWSN:
The hybrid underwater sensor architecture consists of underwater robots and sensor nodes which collaboratively work for completing application specific task and thus named as hybrid architecture. It includes costly autonomous vehicles (AUV, ROV) and ordinary sensor nodes which can be stationery or mobile. In hybrid architecture, the AUV is send to each sensor node for collecting the data and then AUV come backs to a preprogrammed location where the AUV offloads the data. However, all these solutions are still inadequate to provide a network architecture which is energy-efficient and adaptive with the changes of topology with time. Therefore, "architecture of UWSN" is open research issue which needs to be investigated further by research community.

The latter classification of UWSN architectures is based on spatial coverage property which include one-dimensional, two-dimensional, three-dimensional, and four-dimensional UWSN architecture. Note that before going to this classification remember that "Gateway" written in text means buoys, whereas "surface station" or "sink node" means floating ship on water surface and the "anchored sensor nodes" are the sensors deployed at bottom of ocean. Also note that surface station and onshore station are two different things. The onshore station is a stand-alone building with long signal towers.

1.4.1 One-Dimensional UWSN Architecture (1D-UWSN)

In one-dimensional architecture, the single sensor node is a network. Every sensor is self-governing and is alone responsible to detect and transmit data to the remote station. The example of 1D-UWSN can be a self-ruling submersible vehicle AUV which sense and collect data about submerged properties and then transmit data to the remote station.

1.4.2 Two-Dimensional UWSN Architecture (2D-UWSN)

In the two-dimensional UWSN architecture, all sensor nodes are assumed to be at the same depth. For example, nodes may be deployed on the ocean surface or at the bottom of ocean, or they may be floating at a certain depth. Mostly all the sensors at bottom of ocean are connected to a gateway node (i.e. buoy) by means of wireless acoustic links. But one can also use optical and radio links as communication medium.

The diagrammatic representation in Fig. 1.2 shows that the key characteristic of 2D architecture is that the sensors at ocean floor can be organized into clusters and then interconnected to one or more buoys, also known as underwater gateways. These gateways then relay data from ocean bottom network to surface sink node which is generally the ship on water surface. Therefore, the communication from sensor to gateway node and then from gateway node to sink node is said as two-dimensional communication. The vertical and horizontal transceivers are used for communication to-and-from gateways.

Vertical Transceivers:
The vertical transceiver has long range which is used by gateway node to relay data to a sink node which is a surface ship. Thus, practically the complete gathered data present in buoy is send to the ship on water surface using vertical transceivers.

Horizontal Transceivers:
The gateway node uses horizontal transceiver for communicating with anchored sensor nodes. The communication is two-directional, (i) sending commands and configuration information from gateway to anchored sensor nodes and (ii) sending the sensed data from anchored sensor nodes to gateway.

The surface station is also equipped with two transceiver, acoustic and radio transceiver. The acoustic transceiver handles multiple communications coming from gateways (buoys), whereas the radio transmitter or satellite transmitter is used to communicate with onshore station.

The clustering of anchored sensor nodes is optional part in 2D-UWSN architecture. It completely depends on the underwater application for which the UWSN is being developed and on the architecture developer. But to reduce the routing

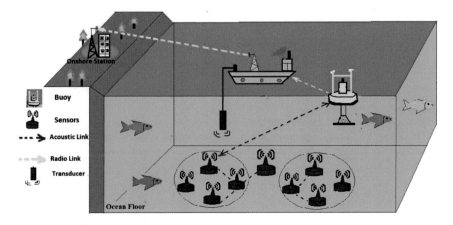

Fig. 1.2 Two-dimensional UWSN architecture

complexity and thus the energy consumption of sensor node, usage of clustering architecture is a feasible option. The possibilities other than clustering are that each sensor can be directly connected to sink node via direct links or through multi-hop paths.

The simplest way of networking is to make each sensor directly send its gathered data to the surface station. But this idea fails in terms of energy efficiency and data reliability, since the distance between sensor node and surface station is very long and thus energy required for data transmission by sensor node is also increased.

The next method is by using multi-hop paths, in which data generated from each sensor is passed through multiple intermediate sensor nodes, until it reaches the surface station. Thus, multi-hop architecture is energy efficient and confirms data reliability but at same time it increases routing complexity. Therefore, to design 2D-UWSN architecture which is energy efficient and whose underwater paths are having minimum signalling overhead is still a research challenge. The applications of 2D-UWSN architectures may be in environmental monitoring systems.

1.4.3   Three-Dimensional UWSN Architecture (3D-UWSN)

In the three-dimensional UWSN, each sensor node may be floating at an arbitrary depth to gather data from different depths. The 3D-UWSN architecture is used in project where the bottom sensor nodes are not adequate for detecting and sensing data in three-dimensional space. Thus, in 3D-UWSN architecture the autonomous and unmanned underwater vehicles (AUVs and UUVs) are also used along with the primary sensor node. The 3D-UWSN architecture constitutes the fixed portion of anchor sensor nodes and floating area covered by autonomous vehicles. The architecture is useful in underwater surveillance application and in monitoring of oceanographic environment.

There are many challenges in 3D-UWSN architecture design, because of the sensor nodes mobility due to ocean currents. To address these challenges, there are two solutions which are as follows:

(i) The first possible solution is to attach each sensor to the gateway node (buoy), by using wired connection. The depth of sensors can be changed by adjusting the length of wire. This solution enables easy and faster deployment of sensors, but at same time imagine the situation of multiple underwater sensors attached to multiple buoys which are floating on water surface, thus obstructing ships navigation path. Also, the sensors deployed in such way are harmful in underwater military application as these nodes are easily visible to enemies for deactivating sensors or to modify sensor hardware.

(ii) The other solution is to anchor each sensor node to the bottom of ocean. The anchored sensor node is also attached to a floating buoy that can be inflated by a pump. The job of buoy is to pull the sensor towards ocean surface in order to stabilize the depth of sensor node. The depth of sensors can be regulated by adjusting the length of wire that connects the sensor to the anchor by means of an electronic engine that resides on sensor.

The pictorial representation of 3D-UWSN architecture is as shown in Fig. 1.3, where the anchored sensor nodes cover the entire monitoring region and then AUV collects data from each senor node. After gathering data, the AUV come backs to preprogrammed location on water surface from where the data is offloaded to surface station.

1.4.4 Four-Dimensional UWSN Architecture (4D-UWSN)

If carefully observed the four-dimensional representation in Fig. 1.4, it can be said that 4D-UWSN architecture is a combination of three-dimensional and two-dimensional sensor architecture as it consists of floating and fixed sensor nodes. The advanced and additional feature in four-dimensional architecture is using remotely operated vehicle (ROVs) as the intermediate node to relay data. The ROV is used as floating sensor node which collects and relay data from each sensor node or cluster head to the gateway node. The gateway node then sends the data to an onshore station. The data transmission situation among ROV and sensor node depends on distance, size of information and the communication medium acoustic or radio. The data transmission in which the sensor has substantial information to transmit and are near to ROVs can utilize radio connections, whereas the sensors which have less information and are at long distance from ROV can use acoustic links.

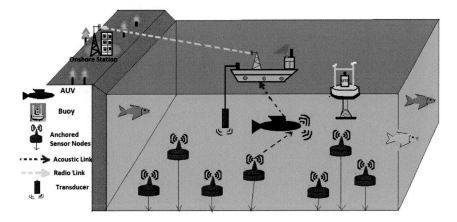

Fig. 1.3 Three-dimensional UWSN architecture

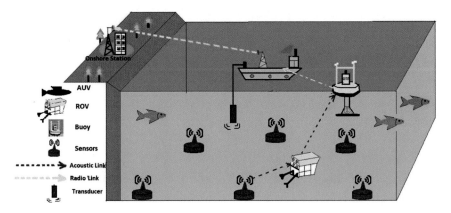

Fig. 1.4 Four-dimensional UWSN architecture

Therefore, based on these basic UWSN architectures, the research community has come up with many different architectures. For example, the Aqua-Net, Tic-Tac Toe Arch, Trees of Wheel (ToW) and so on. All these different architectures are detailed explained in third chapter of the book. There are many aspects while designing an architecture like total number of sensors required, location of sensor deployment, method of deployment, etc. All this aspect is also covered in the chapter "Study of Different Architectural Designs for UWSN".

References

1. Hanson F, Radic S (2008) High bandwidth underwater optical communication. Optical Society of America. Appl Opt 47(2):277–283
2. Nakamura K, Mizukoshi I, Hanawa M (2015) Optical wireless transmission of 405 nm, 1.45 Gbit/s optical IM/DD-OFDM signals through a 4.8 m underwater channel. Optical society of America. Opt Express 23(2):1558–1566
3. Oubei HM, Duran JR et al (2015) 4.8 Gbit/s 16-QAM-OFDM transmission based on compact 450-mm laser for underwater wireless optical communication. Optical Society of America. Opt Express 23(18):23302–23309
4. Wu TC, Chi YC, Wang HY, Tsai CT, Lin GR (2017) Blue laser diode enables underwater communication at 12.4 Gbps. Scientific Reports 2017
5. Sonardyne, BlueComm 200, Optical Sensor. https://www.sonardyne.com/product/bluecomm-underwater-optical-communication-system/. 26 Apr 2020
6. National Oceanic and Atmospheric Administration. https://www.noaa.gov/. 26 Apr 2020
7. Woollett RS (1970) Ultrasonic transducer. Ultrasonics 8(4):243–253, Oct 1970
8. Ahmed KA et al (2017) Design of polyimide based piezoelectric micromachined ultrasonic transducer for underwater imaging application. In: Proceedings of the international conference on imaging, signal processing and communication, pp 63–66, 26–28 July 2017
9. Zhang QQ et al (2006) High frequency broadband PZT thick film ultrasonic transducers for medical imaging applications. Ultrasonic 44:e711–e715. Elsevier

10. Boloni L et al (2013) Scheduling data transmission of underwater sensor nodes for maximizing value of information. In: 2013 IEEE global communications conference (GLOBECOM), pp 438–443. IEEE
11. Dubai municipality utilizes YSI real-time water quality monitoring instrumentation
12. Guan J, Huang J, Lu J, Wang J (2013) Topology structure design for UWSN. In: 2013 IEEE international conference of IEEE Region 10 (TENCON 2013), pp 1–4. IEEE, 2013

Chapter 2
Communication Mediums in UWSN

The underwater sensor network (USN) is used in various domains from surveillance to different monitoring applications like monitoring of gas and oil pipelines, aquatic species or water quality. The usage and increasing demand of UWSN in different fields shows the significance of UWSN and its communication medium. In general, the connectivity in UWSN can be broadly classified into two types Wired and Wireless underwater sensor networks. In wired UWSN, the coper wires or optical fibre cables are used as connecting mediums between sensors whereas acoustic waves, light beams or low-frequency radio waves are used as communication mediums in wireless underwater sensor network. The communication mediums are soul of any sensor networks, and thus, they are required to be deeply studied and wisely used during architectural designing of sensor networks. Hence, in this chapter we will study all possible communication mediums used in underwater sensor network.

2.1 Wired Underwater Sensor Network (W-USN)

The traditional approach of wired connection between underwaters sensors can be used for applications which are limited to small monitoring areas or which requires more power supply for functioning. The wiring is also mostly used in applications which requires special purpose underwater sensors that measure only specific attributes like temperature, pressure, pH contains and many other important parameters. Wired connection is a multipurpose communication medium, as used in data transmission and secondly the electrical power transmission for continuous functioning of underwater sensors. The W-USN is easy to deploy and also provides electrical supply to sensors through network wires. However, the W-USN has various problems like the physical damage to network wire may lead to partial or complete damage of underwater sensor network. The W-USN is more prone to security threats as the attacker can easily block the network by cutting some specific

P. N. Mahalle et al., *The Underwater World for Digital Data Transmission*,
SpringerBriefs in Computational Intelligence,
https://doi.org/10.1007/978-981-16-1307-4_2

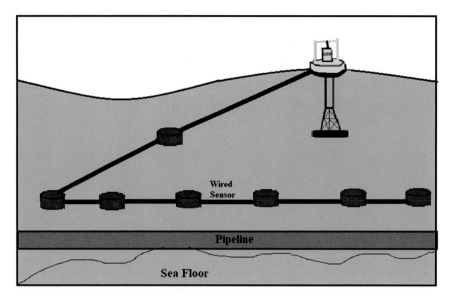

Fig. 2.1 Wired underwater sensor network (W-USN)

numbers of network wires. Thus, any intentional or natural physical wire damage, security or reliability problem face by W-USN make it an unfeasible connectivity solution for underwater sensor network. The W-USN is mostly used for monitoring of gas pipeline lead below water as depicted in Fig. 2.1.

2.2 Underwater Wireless Sensor Network (UWSN)

The wirelessly connected underwater network is popularly known as UWSN, i.e. underwater wireless sensor network as shown in Fig. 2.2. The UWSN is more reliable and secure as compared to the W-USN as wireless network can continue functioning even if some nodes are faulty or intentionally disabled. There are mainly three communication mediums used for wireless communication, acoustic, optical and radio waves. In UWSN, generally long-range acoustic waves or dense deployment of sensor nodes can be done for transmitting sensed information to its destination and to maintain the network connectivity in existence of faulty nodes. The optical and radio communication mediums are less used as compared to acoustic wave. However, the underwater data transmission done using acoustic wave as wireless communication medium pose major challenges like limited bandwidth, propagation delay, absorption, attenuation and many more. Each challenge is detailed explained in subsection, "Challenges in UWSN".

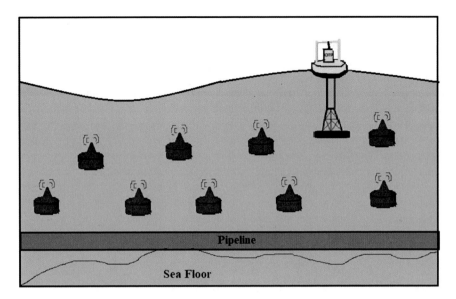

Fig. 2.2 Wireless underwater sensor network

In W-USN, the wires are multipurpose as it transfers data along with electrical power, so energy of sensor node is not a constrained in wired underwater sensor network. But power is a main challenge for wireless underwater networks, and thus, the network designer and protocol developers have to consider energy of sensor node as a precious resource in UWSN. The sensors of wireless network consume highest amount of energy and memory in data transmission, and thus, careful resource management is the key for extending life of wireless underwater sensor network with minimum use of sensor battery, memory and processor. But balancing of this resource with secure data transmission in underwater wireless sensor network is difficult to achieve and hence network availability is one of the major problems in wireless network as compared to wired underwater networks.

There are many underwater applications which use UWSN; one of the examples is monitoring vibrations of gas pipeline [1]. These pipelines are deployed at bottom of ocean where the surface of seafloor is uneven due to which some parts of pipeline is rested on seafloor, while some part is lifted up from the seafloor hence leaving a gap between pipeline and seafloor. The strong sea currents may induce high vibrations in these free gaps, which indirectly introduces a high pressure on the parts of pipeline. Therefore, underwater wireless sensor network was used to monitor the vibrations on pipeline. The network also used remotely operated vehicle (ROV) to replace the disable nodes. The maintenance cost of such monitoring networks is more due to need of periodic replacement of sensor whose battery life is only for six months.

2.3 Types of Wireless Communication Mediums in UWSN

The different wave types used as communication mediums for communicating wirelessly in underwater sensor network are radio-frequency waves, acoustic waves and optical waves. Each communication medium has some advantages and disadvantages. Therefore, choice of communication channel should be purely done on basis of underwater application to be developed. For example, applications which require reliable data transmission over long distance with acceptable propagation delay and where security is secondary, then one can use acoustic waves as communication channel. However, applications which require faster and secure data transmission over short distance, optical wave is better choice. Radio-frequency waves are mandatory in most of the underwater applications to relay data from the gateway node to onshore station. The radio waves are less used as underwater data communication channel, as radio waves are absorbed in water.

2.3.1 Radio-Frequency Communication

The high- and low-frequency radio waves also known as electromagnetic wave can propagate underwater only at certain frequencies and for limited distance. The high-frequency radio waves cannot penetrate into water due to the phenomena of attenuation and high reflection. The extremely low-frequency radio waves are used for data transmission in underwater. The distance propagated by low-frequency radio waves depends on type of water, for example in pure water radio waves can travel a few meters and in sea water it suffers from severe conduction. In seawater for distance of 200 m, the data rate is 50–100 b/s and for same distance in fresh water, the data rate is 100–200 b/s. The underwater submarine communication by USA and Russia used 76 Hz or 82 Hz extremely low frequencies radio waves which transmitted few characters per minute [2]. The other example is two drivers apart at distance of 1 m and at depth of 2 m directly communicating with onshore station using radio waves as it the low-frequency radios have property of crossing water-to-air boundary. However, the use of extreme low radio frequency requires huge antennas that can consume lots of power. In summary, low-frequency radio waves offer advantages including transmission crossing water and air boundaries and the minimum data rate. But still radio communication is not preferred over acoustic wave due to the drawbacks like absorption and conductivity.

2.3.2 Acoustic Communication

The acoustic wave popularly known as sound waves, is commonly used as communication medium for underwater wireless data transmission. The acoustic communication is used in majority of underwater applications because distance travel by this wave is in kilometres and the speed of acoustic wave in water is four to five times faster than air. However, acoustic communication channel also suffers from limited bandwidth and data rate. The acoustic mode of communication in underwater environment poses various challenges on data transmission, such as multipath propagation, absorption, noise, Doppler spread, propagation delay and path loss. These challenges make the available bandwidth of communication channel limited and dependent on frequency and distance to be travelled. If distance is longer, then bandwidth available is in few kHz and for shorter distance bandwidth available is more than 100 kHz. In numerical form, if distance is 1000 km then bandwidth is less then 1 kHz and if distance is 0.1 km then bandwidth available will be more than 100 kHz. The data rate for both longer and shorter distance is low almost in kbits/sec [3].

2.3.3 Optical Communication

The underwater data transmission done using LED, laser or blue–green lights, FSO are used for optical communication in UWSN. The wireless optical communication medium is advantageous over acoustic and radio waves in case of higher data rates and greater bandwidth for short distance. The bandwidth of RF wave is tens of Hz to GHz, while optical waves can reach up to hundreds of GHz. Also, radio waves require large antennas, more battery for data transmission and highly attenuates in sea water; hence, optical wave is good option for short-distance communication. However, optical communication medium for underwater environment also has to face several challenges due to scattering phenomena or optical beam absorption by water. The theoretical and experimental studies [4–7] done show that optical beam can be used for underwater wireless data transmission with larger bandwidth over shorter distances. The research work for optical wave as communication medium in UWSN was started from 1992, using argon-ion laser [8] which had bandwidth of 50 Mbps over distance of 9 m. Then in 1995, a theoretical study for LED-based underwater optical communication was carried out for distance of 20 and 30 m with speed of 10 and 1 Mbps [9]. The research on unidirectional [10], bidirectional [11] and omnidirectional [12] underwater optical communication link is still going and most of the work had optical link ranging up to 10 m. The high-bandwidth optical waves are influenced by scattering, temperature changes and beam steering. However, the blue–green light suffers from low attenuation as compared to radio and acoustic waves. For this reason, there has been a vast development of blue–green sources and detectors.

2.4 Challenges in UWSN

The each of three communication mediums radio, acoustic and optical waves face many challenges in propagation due to the uncommon physiochemical underwater environment. The water has chemical properties like salinity, pH levels, alkalinity, presence of ions and molecules, dissolved gasses and physical properties like thermal conductivity, turbidity, density, transparency. All these physiochemical properties of water affect the waves propagation and thus indirectly the underwater data communication. The other important factor of underwater networks is that sensors replacement is difficult and battery is a precocious resource in UWSN. The various aspects within UWSN like distance between sensor node, localization of nodes, synchronization among nodes, underwater data transmission protocols, security algorithms, energy consumption of nodes, UWSN architectures and many more aspects must be designed and developed by considering the list of challenges in UWSN. The influencing challenges like propagation delay, multipath effect, path loss, Doppler spread, scattering, energy consumption, noise make sever effect on available bandwidth of communication channel. Because the influencing factors over communication mediums make channel bandwidth limited and distance-frequency dependent.

2.4.1 Propagation Delay

The amount of time taken by wave to reach its destination is said as propagation speed. Then propagation delay is increased in amount of time required for wave head to reach its desire destination over underwater medium. The increased time is considered as a delay, which occurs mainly due to water and wave properties. The increase in propagation distance due to the spreading of wave front also causes propagation delay. The larger propagation delays are harmful for protocol designs and also reduces the overall throughput of system. The varying and increased propagation delays can be disastrous for real-time underwater applications like underwater tsunami detection system.

2.4.2 Limited Bandwidth

Each of the communication medium used for UWSN has a different available bandwidth. The bandwidth of underwater communication channel is significantly affected by influencing factors of underwater communication channel like path loss, scattering, Doppler spread, noise and water temperature. The distance travelled by acoustic wave is generally divided into four categories longest (1000 km), long (10–100 km), moderate (1–10 km), short (0.1–1 km) and shortest (less than

0.1 km) distance. The available bandwidth for acoustic channel in each category is different as it is distance dependent. For example, the bandwidth for longest distance is less than 1 kHz and for long it's up to 5 kHz, similarly for short distance bandwidth is 20–50 kHz and for shortest it is more than 100 kHz. In summary, for acoustic channel shorter the distance larger is the available bandwidth. The same principle also applies for optical channel but the bandwidth available is in MHz or GHz. Therefore, optical channel as communication medium is best suitable for short-distance communication up to 10 m. The bandwidth for radio channel is in MHz for very short distance, and hence, the radio waves are least used communication medium for UWSN.

2.4.3 Noise

The noise in a communication channel is differentiated into two types manmade and ambient noise. The noise originated from manmade activity like shipping and machinery noise. The manmade noise is also said as site-specific noise which is limited only for particular areas. The noise generated due to natural phenomena like tides, rain, water movements, tsunami wave generation, underwater earthquakes, bubbles, thermal noise, sound made by aquatic animals, etc., all are said as ambient noise or background noise. Both types of noise directly affect the SNR of receivers, and hence, noise is a critical issue which should be consider when selecting communication medium for UWSN. The frequency of sound is used to differentiate source of ambient noise, as shown in Table 2.1.

2.4.4 Doppler Spread

The dynamic underwater sensor nodes float with ocean currents, hence resulting into motion of transmitter and receiver, which results in Doppler spread. The Doppler spread produces a frequency conversion which is denoted as $f = (1 + \Delta v/c)\ f0$, where f is Doppler effect and $f0$ is actual frequency. Therefore, in Doppler spread it can be seen that actual frequency is no longer same but is basically increased. The significant Doppler frequency range results into degradation of underwater digital data transmission, and hence, it is important to consider Doppler effect during development of underwater protocol and data-dependent underwater applications.

Table 2.1 Recognition of ambient noise source

Frequency range	Ambient noise source
20–500 Hz	Surface ships
500–100,000 Hz	Bubbles
100 kHz	Brownian motion of water molecules

2.4.5 Path Loss

The path loss mainly occurs in acoustic channel due to three reasons, (i) the amplitude of acoustic wave diminishes as it travels, (ii) attenuation of acoustic wave and (iii) geometric spread. The sound (acoustic) waves attenuate in water as it travels, due to conversion of sound energy into heat energy. The attenuation increases with increase in frequency and distance travelled, and hence, attenuation is distance and frequency dependent. Other than, conversion of energies attenuation also occurs due to reflection, dispersion, refraction and scattering of radio, acoustic and optical waves. Therefore, attenuation of wave ultimately leads to sever path loss. Another reason of path loss in communication channel, for underwater environment is expansion of wave fronts which is referred as geometric spreading. The geometric spread increases with distance but is not affected by change in frequencies. The two common types of geometric spreading describing propagation losses with increasing distance are, (i) spherical spreading, which occurs in deep water and is omni-directional point source, (ii) cylindrical spreading which occur in shallow water and is only a horizontal radiation. The major reason for path loss is due to spherical spreading.

2.4.6 Multipath Effect

The multipath effect occurs due to reflection of sound waves on water surface or ocean floor, any rough surface like sediments and so on. The reflection of wave occurs due to low speed of water and also the variation in speed of water at different depths. The reflection leads to multipath effect, which then degrades communication signal strength. The multipath effect is dependent on distance between sensor nodes and depth of water. The multipath effect generates intersymbol interference which is also responsible for degradation of communication signal.

2.4.7 Properties of Water

The basic knowledge of chemical and physical properties of water is important to understand the effect of water temperature, depth and frequency on the communication mediums. The temperature and salinity decrease with depth of water, whereas pressure increases with depth. The fundamental properties of water like density decrease with depth and viscosity decreases with temperature but velocity of sound increases with depth. In summary, there are different variations with depth, temperature and frequency. The study of water properties helps in reducing the challenges faced by underwater communication mediums.

2.4.8 Energy Constraints

The battery power is the precious resource of UWSN, as sensor battery is not replaceable in UWSN. The battery is mainly required for data transmission and data computation. The energy consumption in each communication medium is different as acoustic modem requires more energy to transmit data as compare to optical modems.

2.5 Comparative Study of UWSN Communication Mediums

See Table 2.2.

2.6 Future of Underwater Communication Mediums

In this chapter, we had an overview on underwater communication mediums and main challenges faced by communication channel. The advancement in underwater communication techniques is still required for efficient underwater communication required for various enhanced underwater applications. The list of challenging factors also creates high bit error rate due to which the communication channel gets open for active and passive attacks. Therefore, there is a need to overcome each challenge over communication medium so as to develop a secure and efficient underwater network.

Table 2.2 Comparative study of underwater communication mediums

Parameters	Radio waves	Acoustic	Optical
Distance	10 m	Few km	10–100 m
Bandwidth	MHz	Distance dependent	MHz or GHz
Attenuation	Highest	Moderate	Lowest
Bit rate	up to Mbps	in Kbps	up to Gbps
Power requirement	High (few mW to 100 Watts)	High (above 10 Watts)	Low (few Watt)
Latency	Moderate	High	Low
Speed	2.25×10^8 m/s	1500 m/s	2.25×10^8 m/s
Cost	High	High	Low

References

1. Manum M, Schmid M (2007) Monitoring in a harsh environment. In: Computing and control engineering, IEEE, Nov 2007
2. extremely low frequency transmitter site, Clam Lake, Wisconsin, The United States Navy Fact File. https://fas.org/nuke/guide/usa/c3i/fs_clam_lake_elf2003.pdf
3. Captipovic J (1990) Performance limitations in underwater acoustic telemetry. IEEE J Oceanic Eng 15:205–216
4. Lu F et al (2009) Short paper: low-cost medium-range optical underwater modem. In: Proceedings of the fourth ACM international workshop on under water networks, ACM 2009
5. Yi X et al (2015) Underwater optical communication performance for laser beam propagation through weak oceanic turbulence. Appl Opt 54(6):1273–1278. Optical Society of America
6. Tu B et al (2013) Acquisition probability analysis of ultra-wide FOV acquisition scheme in optical links under impact of atmospheric turbulence. Appl Opt 52(14):3147–3155
7. Kaushal H et al (2016) Underwater optical wireless communication. IEEE, Apr 2016
8. Snow JB et al (1992) Underwater propagation of high-data-rate laser communications pulses. SPIE digital Library, Dec 1992
9. Bales JW et al (1995) High-bandwidth, low-power, short-range optical communication underwater. In: International symposium unmanned, untethered submersible technology, Durham, NH, USA
10. Vasilescu I, Kotay K, Rus D, Dunbabin M, Corke P (2005) Data collection, storage, and retrieval with an underwater sensor network. In: 3rd international conference embedded network sensor system, 2005
11. Doniec M, Rus D (2010) Bidirectional optical communication with aqua optical II. In: IEEE international conference communication system, Nov 2010
12. Fairetal N et al (2006) Optical modem technology for seafloor observatories. In: IEEE OCEANS, Boston, MA, USA, Sep 2006

Chapter 3
Protocol Layers

The data exchanged between sensor nodes is defined by a protocol—set of rules for communication. There are many such protocols developed for UWSN with each handling a particular aspect of communication. These protocols are generally structured together to form a stack commonly known as protocol layers. As studied in second chapter, radio waves cannot travel long distance into water which makes the UWSN different from TWSN and hence their protocol layers. The protocol stack for UWSN nodes is of four layers, as shown in Fig. 3.1. The application layer is not present in underwater network architecture as it is directly the client machine situated at off-shore station. Each layer in network makes easy classification of protocols developed for UWSN and also defines the service to be implemented by each layer.

In state of the art, various protocols are designed for routing protocols, MAC protocols, connection-oriented and connection-less protocols, node synchronization protocol and many more but UWSN protocol stack currently does not include any strong recommendations on development of protocol for underwater data authentication and encryption. The reason for defining protocol layers is standardization of UWSN [1] to provide exchange and use of information among sensor nodes from different manufacturers and researchers and second reason is to accelerate research in the challenging field of UWSN. Thus, by setting standards on nodes internal protocol layer will allow cross-layer optimization, portability of software within protocol layers, integration of applications with various ranges of sensor nodes. The research community will also be greatly beneficial due to standardization of protocol layers. For example, the standardization in protocols to access the gateway node from terrestrial networks will allow applications to access underwater nodes easily [2]. The end-users can also be given direct access to lower layers functionality if the acoustic modem suppliers share their applications APIs to user [3]. Figure 3.2 shows the suggested interfaces for message transmission among all layer, with REQ, RSP and NTF messages send among the layers. The underwater network architecture includes an additional framework application programming interface (FAPI) for various reasons like to increase the portability of layers and to

© The Author(s), under exclusive license to Springer Nature Singapore Pte Ltd. 2021   33
P. N. Mahalle et al., *The Underwater World for Digital Data Transmission*,
SpringerBriefs in Computational Intelligence,
https://doi.org/10.1007/978-981-16-1307-4_3

Fig. 3.1 Protocol layers

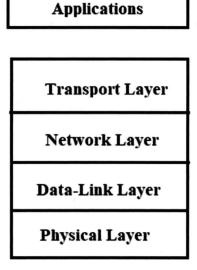

hide operating systems and hardware functionalities. The above layers are usually said as higher layer the transport layer is middleware and below layers (physical and data link) are said as lower layer. The top layers of protocol stack invoke the bottom layers via a REQ message and in response to it the middleware or lower layer sends back RSP message. The ERR RSP message with error codes is used to report errors. And the NTF message is used if layer is not granted verifiable

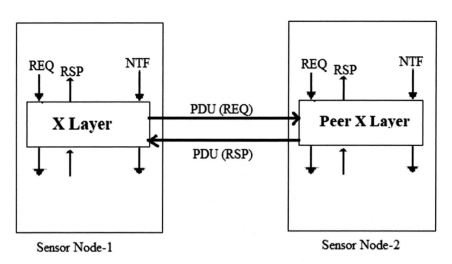

Fig. 3.2 Underwater network architecture layer interface

permission to send or receive messages. The protocol data units (PDUs) are used for logical communication between one layer to the peer layer. This type of communication is symmetric, and hence, layer may send a PDU (REQ) message to its peer layer and can get response message as PDU (RSP).

In recent years, an increasing graph of research in the domain of UWSN is observed due to its usage in defence, oil and gas industries, oceanographic studies and monitoring applications. In the literature, there are various UWSN physical layer protocols, routing protocols and large number of medium access control protocols are surveyed and proposed. In this third chapter, we will cover the basics of all four layers with comparative studies of their protocols.

In summary, the physical layer aims for proper utilization of bandwidth, whereas the data-link layers main objective is to reduce energy consumption of nodes. The network layer is responsible to deal with dynamic topology of UWSN, the dead nodes and shadow zones. From state of the art, the research on protocol layers is still in its infancy and requires more mathematical-based research and simulations to be done for underwater physical, data link, network and transport layer.

3.1 Physical Layer

The lowest layer of protocol stack, known as physical layer, is responsible for sending data from one underwater modem to another. The physical layer is not concerned with meaning of data and only deal with physical connection among modems by using underwater communication mediums. The main job of layer is to only receive and transport the signals to its above layers. The devices that operate at physical layer are switches, controller, memory, sensors and interface between switch and controller as shown in Fig. 3.3 where the switches are used to transmit data according to the flow table and controllers are used to decide the data forwarding strategy. The controller receives data from the underwater modem and stores it on the memory chip. The stored data is then processed by controller/CPU and then sends or receives data packets to-and-from the peer underwater modems.

The physical layer is also responsible for modulation (FSK, PSK, QAM, OFDM), demodulation, multi-carrier multiplexing, channel equalization, channel coding, phase conjugation, spatial modulation, collaborative communication and error correction. In 90s, the acoustic modem was developed using frequency shift keying (FSK) which was non-coherent modulation technique. Then for long-distance coverage the coherent modulation techniques such as phase shift keying (PSK) and quadrature amplitude modulation (QAM) came into use. These modulation techniques also had various drawbacks like Doppler spread, low data rate, increased propagation delays. Therefore, other technique named orthogonal frequency division multiplexing (OFDM) modulation was used for achieving high data rate for long-distances which is challenging to develop due to low-available bandwidth of acoustic links.

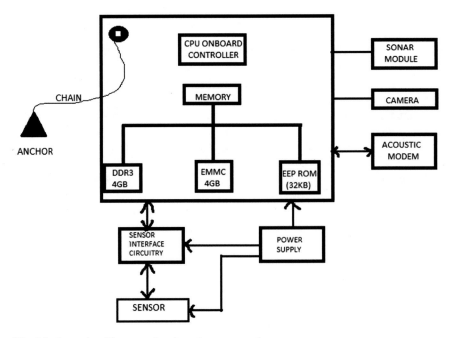

Fig. 3.3 Internal architecture of underwater sensor node

In a study, software-defined radio (SDR) [4] is used to construct the physical layer based on software-defined network (SDN) [4]. The SDR was used to perform modulation, demodulation, channel coding, source coding and multiple-input multiple-output (MIMO) precoding. The authors have designed and implemented various different modulations and encoding techniques which are embedded into physical layer developed using SDN. The SDR provided higher flexibility by allowing encapsulation of low- and high-speed underwater communication.

The standardization in protocol layers is one of the important aspects that will allow the physical layer implementation to enable different technologies like PSK, OFDM, TDMA, CDMA with to allow compatibility and interoperability with modems from different vendors. The Benthos, the commercial modem makers [5], developed compatible underwater modems after the authors Freitag [6] published micromodem as the first standards. The standardization in UWSN is still in its infancy, not due to lack in technology but due to lack in research and development. The standardization in UWSN will lead to easy set-up of UWSN just same as setting up of TWSN, and users will be opened to all types of modems from different manufacturers. The targeted interfaces in UWSN which should be standardized are shown with the help of Fig. 3.4.

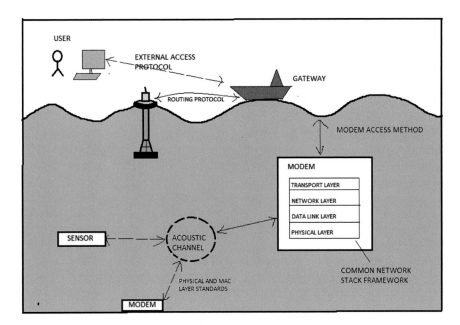

Fig. 3.4 Interfaces for standardization

3.2 Data-Link Layer

The basic responsibility of data-link layer is to provide reliability service to its higher layers, and other is to divide data received from network layer into data frame in order to provide service to its lower layer (physical layer); in additional, it also works on error detection, for example, CRC check. In previous chapters, we have seen that the TWSN is different from UWSN, due to various reasons like low data rates, higher propagation delays and higher bit error rate. The research work in [7] has showcased many drawbacks of using existing TWSN MAC protocol for UWSN. Hence, its mandatory to design a new MAC protocol, to achieve node co-ordination in UWSN. There are two types of MAC protocol, connection-based and connection-less MAC protocols. The nodes with connection-based protocols like ALOHA, SALOHA, CSMA, medium access collision avoidance MACA, MACAW [8, 9], compete to share the communication medium, whereas the traditional connection-less MAC protocols like TDMA, FDMA, code division multiple access (CDMA), space division multiple access (SDMA) are used for collision avoidance. The connection-based protocol used messages like request-to-send RTS, clear-to-send CTS, DATA, acknowledgment ACK, automatic repeat request ARQ sequences. As compared to scheduled protocols in Seaweb project [10], these connection-based protocols are more effective, but for some networks, these RTS/CTS can also be inefficient [11]. In research study, it was observed that power control and adaptive modulation are the two main factors for increasing channel capacity and efficiency.

Pure ALOHA: The ACK sequence is used by receiver to acknowledge that data is received. If sender in case does not receive the ACK, then the packets are retransmitted by sender. The connection-based S-ALOHA version of pure ALOHA was developed to increase the efficiency by using discrete timeslots, thus reducing collision and increasing throughput. The more advance version, where ALOHA-AN (Advance Notification) and ALOHA-CA (Collision Avoidance) with small size of advanced notification packets, helps to build the database tables for each node. The database tables are used by node to verify the collision avoidance at neighbouring nodes.

Medium access collision avoidance (MACA): It is connection-based protocol which use RTS, CTS handshaking. The MACAW is extension of MACA protocol which includes RTS, CTS, DATA, ACK packets as they have advantage over communication medium. This protocol is highly used in situations where a number of nodes required are more and node are not synchronized [12].

Floor acquisition multiple access (FAMA): The connection-less protocol increases the duration of CTS and RTS packets. Hence at sender and receiver sides, the connection is managed before sending data packets. The FAMA is very unsuitable for UWSN due to various drawbacks like difficult to configure the system after arrival of new node. The enhanced FAMA protocol, FAMA-CF (collision free) with slotting can be used only for centralized topology.

Both types of protocols connection-based and connection-free are used in many UWSN at today's date. The selection of protocol is based on the requirements like time criticality of data, total number of expected nodes, delay tolerance, reliability, sleep–wakeup schedules, network traffic, node mobility and time synchronization. The accurate simulations are still required to work on many different ideas for underwater data-link layer and MAC protocols. The reduction in energy consumption of nodes is also considered as an aspect of data-link layer in recent studies but the literature states that energy efficiency is achieved by compromising data rate.

3.3   Network Layer

The routing of data packets and designing network protocols for UWSN is challenging task due to various drawbacks of underwater networks like variable delays, limited bandwidth, unreliable data transmission, higher BER, node mobility, node failure due to corrosion and multi-path fading. Therefore, the routing protocol used in TWSN cannot be used for UWSN due to a huge difference between both networks. The literature study [13–15] shows that it is not possible one single routing strategy to all underwater applications because each strategy has some pros and cons making it applicable to only specific applications. The routing protocols can be classified according to application scenarios using single-sink, multi-sink and

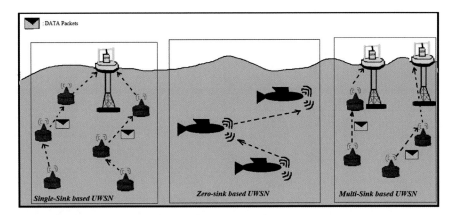

Fig. 3.5 Sink-based UWSN

zero-sink as depicted in Fig. 3.5. The total number of sink node and topology of UWSN are two major factors affecting UWSN. The multi-sink-based UWSN is most preferred, as it improves the routing efficiency by minimizing the transmission range. The single-sink underwater networks also simplify the designing of routing protocols but with zero-sink-based UWSN, it is difficult to design routing protocols.

There are seven routing strategies used for these applications scenarios, which are geo-based routing, source routing, hop-by-hop routing, clustered routing reinforcement learning, cross-layer and opportunistic routing [16].

Geo-based routing: The routing strategy uses the location information of underwater sensor nodes in order to find the best possible route from sender to receiver. The signal strength and GPS are used to find location information of node. But GPS does not work underwater, and signal strength is affected by noise and attenuation; therefore, the location coordinates may not be corrected.

Source routing: The partial or complete planned transmission path and identities of all intermediate nodes is already embedded in the packet sent by sender node but the source routing at same time increases network overhead and routing cost.

Hop-by-hop routing: In this routing scheme, the relay node is responsible to decide its next hop. Hence, there is no restriction on total number of nodes to be used or addition of new node, but all routing path selected may not be the best possible path.

Clustered routing: In clustering, underwater nodes are divided to form clusters but cluster head (gateway node of group selection) is again an issue. The advantageous part is that it reduces data redundancy.

Cross-layer routing: In this strategy, the next transmission node is selected using distance to sink, buffer size, different delays in packet delivery and channel usage. It considers the functions and data of other layers also. The cross-layer routing increases the network performance and energy efficiency.

Table 3.1 Mapping of routing strategies to application scenarios

Application scenario	Routing strategy
All scenarios	Geo-based routing
Static single-sink-based UWSN	Source routing
All scenarios except zero-sink-based UWSN	Hop-by-hop routing
Single-sink-based UWSN	Clustered routing
Static single-sink-based UWSN and zero-sink-based UWSN	Cross-layer routing
Single sink-based UWSN	Reinforcement routing
Mobile multi-sink-based UWSN	Opportunistic routing

Reinforcement learning-based routing: The iterative calculation of reinforcement function is done to select the next hop. Such strategies increase network lifetime but requires powerful modems.

Opportunistic routing: The one single node selects multiple neighbouring nodes as a candidate of next relay nodes and thus ensures reliable data transmission.

Table 3.1 will explain for which application scenario which routing strategy should be used.

3.4 Transport Layer

The transport deals with reliable data transmission which is actually a problem for UWSN. In literature, very less research is done on transport layer as compared to other three layers of protocol stack. The transport layer is mainly responsible for data transmission using two-mode connection-oriented and datagram [17]. The connection-oriented mode mostly does reliable transmission, but datagram mode can do reliable or unreliable delivery of data packets. The segmented data reliable transfer (SDTR) [18] protocol is developed for reliable data transmission in UWSN.

In this third chapter, we have discussed all the four layers of protocol stack. In summary, we can say lots of research work which is still required in this area of UWSN. In physical layer, research about increasing bandwidth of acoustic link, improving quality of underwater communication channel is still in-sufficient. In data-link layer, various MAC protocols are proposed in literature but protocol with more efficient use of sensor battery can still not reached to level of standardization. In network layer, research on routing protocols for mobile UWSN is less focused. Also, the routing should deal with error minimization, node failure and shadow zone [19, 20]. Hence, more and more mathematical-based research and simulations are required to verify the proposed ideas and protocols of each layer.

References

1. Chitre M et al (2008) Underwater acoustic communications and networking: recent advances and future challenges. Springer
2. Stokey RP, Freitag LE, Grund MD (2005) A compact control language for AUV acoustic communication. In: IEEE OCEANS conference 2005
3. Chitre M et al (2006) An architecture for underwater networks. In: IEEE proceedings of the oceans Asia Pacific 2006, Singapore, May 2006
4. Wang J et al (2017) Design of underwater acoustic sensor communication systems based on software-defined networks in Big Data. Int J Distrib Sens Netw 13(7)
5. Freitag L et al (2005) Multi-band acoustic modem for the communications and navigation aid AUV. In: Proceedings of OCEANS 2005 MTS/IEEE, pp 1086–1092
6. Freitag L et al (2005) The WHOI micro-modem: an acoustic communication and navigation system for multiple platforms. In: Proceedings of IEEE Oceans Europe, Brest France, June 2005
7. Xiao Y et al (2011) Performance analysis of ALOHA and p-persistent ALOHA for multi-hop underwater acoustic sensor networks. In: Cluster computation 2011
8. Akyildiz F et al (2002) Underwater acoustic sensor networks: research challenges. Ad Hoc Netw 3:257–279. Elsevier
9. Sozer EM et al (2000) Underwater acoustic networks. IEEE J Ocean Eng 25(1):72–83
10. Rice J et al (2000) Evolution of seaweb underwater acoustic networking. In: OCEANS 2000 MTS/IEEE conference and exhibition
11. Turgay K et al (2006) A mac protocol for tactical underwater surveillance networks. In: Military communications conference, MILCOM 2006. IEEE
12. Kebkal A et al (2005) Data-link protocol for underwater acoustic networks. In: Proceedings of the IEEE Oceans Europe 2005, pp 1174–1180
13. Li N et al (2016) A survey on underwater acoustic sensor network routing protocols. Sensors 16(3):414. https://doi.org/10.3390/s16030414
14. Akyildiz F, Pompili D, Melodia T et al (2006) State-of-the-art in protocol research for underwater acoustic sensor networks. ACM SIGMOBILE Mobile Comput Commun Rev 11(4):7–16
15. Muhammad A et al (2011) A survey on routing techniques in underwater wireless sensor networks. J Netw Comput Appl 34(6):1908–1927
16. Lu Q et al (2017) Routing protocols for underwater acoustic sensor networks: a survey from an application perspective. In: Advances in underwater acoustic, Intech Open
17. Potter J (2006) An architecture for underwater networks. In: IEEE OCEANS conference 2006
18. Peng X et al (2010) SDRT: a reliable data transport protocol for underwater sensor networks. Ad Hoc Netw 8(7):708–722. Elsevier
19. Shashaj A et al (2014) Energy efficient interference-aware routing and scheduling in underwater sensor networks. In: Proceedings of MTS/IEEE OCEANS 2014, 14–19 Sep 2014. St. John's, Canada, pp 1–8
20. Basagni S et al (2012) Channel-aware routing for underwater wireless networks. In: IEEE OCEANS conference 2012

Chapter 4
Threats and Attacks in UWSN

4.1 Information Security and UWSN

Data security and wireless communications are the two wheels of same vehicle. In today's era, data is useless if received from any unsecured wireless network may be terrestrial or underwater. In last 30 years, much of research work has been done in the domain of UWSN starting from underwater sensors and robotics developments, to routing and MAC protocols. In state of the art, the topics which seem to be least focused are security of underwater networks, alternative for acoustic underwater communication medium and hybrid network architectures. Among these three areas, data security for underwater networks is a very delicate issue. The challenges of underwater communication mediums (discussed in second chapter) might be one reason due to which researchers have always given information security, a lowest priority. Hence, previously the data packets do not include any encryption or decryption code, leaving the underwater networks unattended and exposed to various attacks endangering the legitimate data packets. Then growing use of UWSN by oil and gas industries, government agencies for disaster monitoring, and surveillance operations, showed the need to equally prioritized information security of underwater networks, and it has also gained momentum along with development of localization schemes, time synchronization of sensors, standardization of protocol stack, designing of routing strategies, hardware and so on. In fact, data security is of paramount importance and is not replaceable by use of any advanced networking solutions, rather one has to implement security algorithms and protocols to develop a full-fledged networking solution.

In recent studies, researchers have focused on identification of various electronic attacks on UWSN like wormholes, jamming, sinkhole, spoofing, eavesdropping and so on. These attacks are carried out mainly with two intentions, (i) the different underwater applications generate huge confidential data and (ii) the recent standardization of UWSN (discussed in third chapter) have made an easy way for attackers to develop attacking models or strategies for underwater networks.

P. N. Mahalle et al., *The Underwater World for Digital Data Transmission*,
SpringerBriefs in Computational Intelligence,
https://doi.org/10.1007/978-981-16-1307-4_4

Therefore, its need of time to exclusively tailor the security techniques before deployment of UWSN. In literature, its observed that most of the data security techniques of TWSN cannot be used directly for UWSN except the elliptic-curve cryptography [1]. Therefore, an innovative approach or re-thinking on security schemes used for TWSN is required to develop a security solution for underwater networks which can defend attacks by managing the challenges of UWSN.

[Note: The UWSN may also include bidirectional wireless communication between underwater sensors and underwater robots (AUVs or ROVs). Therefore, information security can be implemented as per underwater application communication requirements which may be between two sensors, or between underwater sensor and AUVs or ROVs or vice versa, or sensor to base stations and underwater robots to base station].

4.2   Threats in UWSN

The constantly increasing research interest in the field of underwater data transmission has led to development of large number of underwater applications, starting from oceanographic studies to industrial applications. These happening evolutions in underwater world have increased the importance of underwater data transmission, and thus, data security risk and threats have become more critical issues. The threat is intentional damaging of underwater networks by means of different attacking strategies, which has frightened many underwater application developers and the research community. Hence, in simplicity there is a need to analyse different security threats, to overcome henceforth attacks on underwater networks. The threat analysis discussed with help of below Table 4.1 shows the importance of developing security solution for underwater data transmissions by managing the underwater challenges.

There can be still many numbers of unidentified threats and the corresponding attacks to which UWSN is susceptible. Therefore, developing high-level security algorithms to achieve message authentication, data confidentiality, network integrity, harder security key generation and distribution are some of the preventive measures to make UWSN escape from security attacks.

4.3   Attack Modelling

The threat and attack are two different concepts, and hence, it is required to discuss them separately. In simple terms, threat is like a bad dream in which we only forecast the possibility of something bad happening, whereas the attack is an actual fact which occurs as realising threat. Now, for obvious reason attacks are done where there is something to gain or to damage. In UWSN, the attacker can gain the confidential digital information transmitted over underwater networks with an

Table 4.1 Threat identification and corresponding Problem detection

Threats	Attacks	Problem
Network unavailability	DoS Host-based attack	Intentional disturbance in communication and nodes coordination Network unavailability
Increased energy consumption	Energy exhaustion attack Flooding of packets attack Overwhelming attack	Intentional exhaustion in nodes energy
Modification in legitimate data packet	Message tampering attack	Loss in information confidentiality
Unreliable data transfer	Spoofing, jamming, blackhole, sinkhole, wormhole attacks	Loss of authenticated data Processing of fake information by legitimate node
Hacking of cluster heads	Jamming attack Homing attack	Disrupt entire network and freshness of data Cheap attack with greater damage
Tampering of nodes	Physical attack by capturing an underwater node	Modification in internal memory of node Loss of auditing data
Loss of network integrity	Spoofing-based DoS attack	Production of wrong routing tables Lacks anonymity
Larger propagation delay	Adversary passes message with incorrect path information Wormhole, Sybil, misdirection attack	Lacks identity authentication Larger time period required for intruder detection

intention of damaging the national security of enemy countries. The network where there is nothing to gain or damage then it's an no-attack situation with zero risk. But the increasing applications of UWSN for public health care and national security will always make UWSN as high-risk carrying network. Therefore, to reduce this risk there is need to develop security mechanisms with better understanding of various electronic underwater attacks. Therefore, this subsection gives a detailed explanation of different attacks on UWSN. The attacks on underwater networks are mainly classified into three types depending on site of attack like, protocol layers, attacker nodes and information in transit.

4.3.1   Attacks on Protocol Layers

Figure 4.1 depicts the various attacks launched at different layer of UWSN protocol stack. The attack on any layer affects the flawless functionality of protocol stack; therefore, it is necessary to address layer-wise attack for developing security mechanisms accordingly.

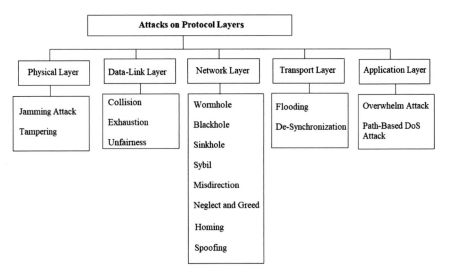

Fig. 4.1 Taxonomy of attacks in protocol layers

Attacks on Physical Layer: The jamming and tampering attack are the two main attacks occurring at physical layer. The physically present sensor nodes and communication link are captured in both jamming and tampering attack but for two different purposes. In jamming attack, the node is captured to continuously send meaningless data or noise signals to neighbouring nodes with an intention of jamming network traffic, whereas tampering includes changing the data present in internal memory of node.

Attacks on Data-Link Layer: The attackers main aim at the MAC layer are related to communication channel like increasing channel access time, increasing propagation delays, degradation in channel quality and so on.

Collision: There are three attacks named collision, exhaustion and unfairness are identified as data-link layer attacks in literature. The multipath effect is main reason for occurrence of collision attack and which ultimately leads to a high error rate.

Exhaustion: In exhaustion attack, the attacker forces a node for retransmission of corrupted data packets with intention of exhausting nodes resource which indirectly leads to reduction of network lifetime.

Unfairness: The forceful unequal distribution of channel access time is known as unfairness attack. The exhaustion and collision attacks may cause unfairness, whereas repeated collision attack causes resource exhaustion.

Attacks on Network Layer: The main service provide by network layer is routing of data packets to its destination through many intermediate nodes. The layer also provides other services like packet sequence control, connection service, flow control, packet switching and so on. Therefore, the targeted area for attackers is network layer services which are base of entire network. There are many attacks in network layer as described in Fig. 4.1.

Wormhole: The attacker creates an attractive high-bandwidth and low-latency wormhole link (tunnel) between two or more malicious nodes [2]. This wormhole link is then used by attacker to create disastrous attacks like complete disruption of routing [3], localization, synchronization services [4]. The link is also used to break the packet sequences, selective packet dropping, packet modification and to launch DoS and man-in-middle attacks.

Blackhole: The blackhole is a malicious node which drops all the packets passing through it. The blackholes are created using different ways like the malicious node broadcast the packet with shortest routing path attracting the network traffic towards it or malicious node creates a scenario showing it as the nearest relay node.

Sinkhole: The sinkhole is the malicious node created near to sink node which prevents the sink node or surface station (buoys) to receive complete sensed data. These is because sinkhole is attacking node with powerful hardware and has capability to drop all or selective data packets.

Sybil: In human beings, we have a disease name multiple-personality disorder, and the same concept is used in Sybil attacks. The attacker creates a single malicious node which shows multiple identities, therefore creating a fake scenario of multiple relay nodes. These identities are newly created by attacker or are spoof from legitimate node.

Misdirection: The attacker misdirects the legitimate node by forwarding packets with wrong route information or wrong routing table information.

Neglect and Greed: The attacker allows passing of acknowledgement packets but intentionally drops or neglects the data packets. The term greedy because the attacker prioritizes his own message. A reduction in hop-count is only the countermeasure for such attacks.

Homing: In this attack, the targets are node with critical responsibilities like cluster heads and sink node. The failure of such nodes will break down the whole network. The countermeasure for such attack is to secure each route of data transmission and to hide important nodes by encryption.

Spoofing: The spoofer (attacker) impersonates the neighbouring nodes that he is a legitimate node. For creation of such scenarios, the attacker creates a fake MAC address and also notes that spoofer has all access rights same as other authorized nodes. The spoofing attack can create network congestion by sending routing tables to interrupting traffic.

Attacks on Transport Layer: The main job in transport layer is to process and transfer network data for use of applications and also prepare application data to transport over network. The other services provided by transport layer are packetization, reliable delivery, connection control, multiplexing and addressing. The two main attacks carried out in transport layer are desynchronization and flooding. In flooding, the attacker targets exhaustion of node resource by flooding the targeted node with connection establishment requests.

The underwater networks can be said as GPS-free networks because GPS is based on electromagnetic waves which do not function underwater. Hence, the time synchronization service at transport layer becomes the most difficult task for underwater networks and thus a hot place for attacker. The messages with wrong sequence numbers are sent to both endpoints leading to desynchronization of data packets.

Attacks on Application Layer: The application layer of UWSN differs a lot from terrestrial networks as there are no standardized protocols for UWSN application layer. The underwater application software are installed directly on surface stations (buoy or surface ship) or at desktop machines at offshore stations. The attacker in these layer tries to break the applications by using different techniques to block data packets coming towards base stations. The two attacks identified in application layer are overwhelm attack and path-based DoS attack. The attacker overwhelms the nodes so as to attract all network traffic towards surface station. The intentional reduction in bandwidth availability, and energy of nodes is known as overwhelm attack. And blocking data coming towards sink nodes by simply consuming all network resources on the path to base station is known as path-based DoS attack.

4.3.2   Attacks Based on Capability of Attacker

The underwater wireless networks have to face many challenges due to presence of water in surrounding environment, and therefore, UWSN also has many constraints like shorten battery life, unreliable data transmission, limited data transmission capacity, less computing power and unreplaceable sensor nodes. These constraints and the broadcasting nature of underwater networks make them vulnerable to many threats and security attacks. There is also a category of attacks which depends on attacker's capability. The word capability in this context means the attacks dependent on strength of attacker like hardware and computational capacity of attacking node or the location of malicious node which can make the attack more disastrous.

Attacks based on Location of Attacker: There can be two locations for an attacker, i.e. internal and external. The internal attack is done by compromising one of the internal legitimate nodes, whereas an external attack is done by using outsider node other than authorized network nodes. In internal attack, the attacker has complete information about internal network and secret keys, but this is not the case with external attacks as they are unaware of cryptographic information. As shown in Fig. 4.2, the internal or external attack can be further classified according to actions (i.e. active or passive) taken by internal or external location of node. The actions taken are said as passive and active attacks.

Passive Attack: The simply monitoring of data transmissions by an attacking node without disrupting any network functionality is known as passive action. Therefore, in passive attacks primary aim is to learn the data traffic, observing the packet flow, identifying cluster heads and its location and to obtain the transmitted data. For example, message relay, impersonation, message distortion and

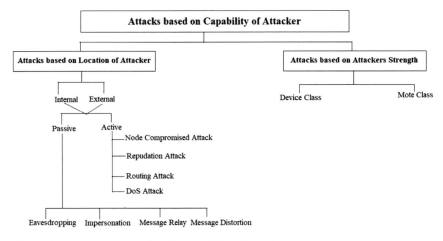

Fig. 4.2   Taxonomy of attacks based on capability of attacker

eavesdropping. The passive attack is most difficult to identify as it is done without harming network functionality.

Active Attack: The attacker modifies, drop, destroy the authorized data packets or may inject fake data packets into network. In active attacks, the network functionality is disturbed by internal or external malicious nodes. The active attack done using inside node cause more damage and is also very difficult to identify. According to attackers aim, the active attacks are further classified as node compromised attack, jamming, repudiation, routing attacks and DoS.

Attacks based on attackers Strength: The device and mote class are two categories of attacks identified based on attacker's strength. In device class, the attacker launches attack by using a specially designed sensor node having high computing power, long battery life, high transmitting power and larger memory. In device class attacks, the attackers' aim is to disguised the legitimate sender, to affect network functionality and to steal secret data. The mote class attack is different from device class attack as the attacker gain access to already present sensor node with an aim to launch an attack. The attacker uses capability of other node to disrupt the network, and hence, such attacks are limited.

4.3.3   Attacks on Information in Transit

In UWSN, data transmission is a biggest aim to develop networks. The wirelessly transmitted data packets contain most of times confidential information depending on underwater application and hence is attracting point for attackers. Therefore, it's important to study various attacks done when data is in transit phase. As shown in

Fig. 4.3, there are mainly five type of attacks identified when information is in transit phase.

Interruption Attack: The attack is done to interrupt the communication link by sending fake messages. To make network services unavailable is the main motto of interruption attack.

Interception Attack: The attacker intercepts the sensor node to gain unauthorized access on network nodes and indirectly on data present in memory chip. This attack occurs in application layer to affect confidentiality of data during its transit phase.

Modification Attack: As the name suggests, capturing the data packets when in transit phase and then modifying the content of data with an aim to misguide the authorized nodes and to affect data integrity.

Fabrication Attack: The attacker injects fake data stream into transiting phase with an objective to lose data authentication.

Replay Attack: The forcibly done repetitions of old data packets again with an intention to exhaust nodes battery is known as replay attack.

4.4 Cross-Layer Approach in UWSN

The 95% of research work done on security algorithms in UWSN deals with only layered security techniques These layered security algorithms only work for one layer. These layer-specific algorithms cannot be said have fully protective techniques as the network suffers from cross-layer attacks. There are various drawbacks of using layered security algorithms for UWSN, like the network resources will soon get exhausted if security mechanisms against an attack are computed separately for each layer. The other drawback of layered security algorithms is the non-adaptive nature to any kind of changes in UWSN like topology changes, node failure and so on. Hence, it's important to develop security system by keeping complete network view, because node energy is precious resource for UWSN.

In practicality, the attackers can launch wormhole attack combined with sinkhole attack with an intention to disrupt the whole network. Such attacks are said as blended attacks [5] which are most disastrous attacks and cannot be handled by routine layered security techniques. Hence, it's necessary to implement cross-layered security techniques like cross-layer authentication, cross-layer key distribution, cross-layer energy management, cross-layer intrusion detection, cross-layer trust model [6].

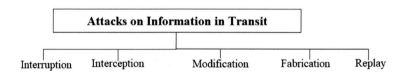

Fig. 4.3 Different attacks occurring during information transit

4.5 Attack-Resistant Strategies for UWSN

The word "resistant" here means to defence against electronic underwater attacks using some strategies in order to protect underwater sensed data and network. The commonly used strategies are message authentication, key agreement protocols, public key cryptography, trust-based security models, intrusion detection systems, cryptographic algorithms and so on. There are also some phases in UWSN like localization, synchronization and routing which should be made secured. In state of the art, for UWSN there are many numbers of security algorithms developed for each strategy. Table 4.2 gives a short description on various security algorithms used and their corresponding defensive attack.

Table 4.2 Study of attack resistant strategies and their defensive attacks

Attack-resistant strategies	Security algorithms	Defensive attack
Authentication schemes	Identity-based authentication Cryptographic suite [7], SecFUN [8], ticket-based authentication [9], low computational complexity authentication scheme [4]	Sybil, misdirection, sinkhole, flooding, fabrication, acknowledgement spoofing
Confidentiality	Underwater jamming detection protocol [10], jamming through analog network coding [10]	Jamming, interruption, interception, modification, tampering
Digital signature schemes	Elliptic curve digital signature algorithm-ECDSA [11], Zhang-Safavi-Naini Susilo-ZSS [12], Boneh-Lynn-Shacham-BLS [13], algebraic signature based protocol [14]	Flooding, desynchronization
Secret key generation and distribution	Iterative key distribution [IKD] [15], RSS-based secret key generation [16], bilinear map-based protocol [14]	Relay attacks, flooding, jamming, spoofing
Trust-based security models	Multi-metric reputation system [17]	Homing
Cryptographic algorithms	CDMA-based analog network coding [18]	Jamming attack
Intrusion detection systems	Dempster–Shafer Theory [DST] [19], context-aware routing [17]	Malicious attack Compromising node
Localization security	Multi-latera privacy preserving localization [20], echo protocol [21], minimum mean square estimation (MMSE)-based location estimation [22]	Information leakage during localization Location spoofing

(continued)

Table 4.2 (continued)

Attack-resistant strategies	Security algorithms	Defensive attack
Secure neighbour discovery phase	Wormhole-resilient secure neighbour discovery [23], SeFLOOD [24]	Wormhole attack, flooding attack
Synchronization security	Cluster-based secure synchronization scheme (CLUSS) [10]	Desynchronization, power exhaustion, collision
Routing security	Secure routing protocol and a set of cryptographic primitives [10], distributed visualization of wormhole [10], wormhole-resilient secure neighbour discovery [10]	Neglect and greed attack. Blackhole, wormhole attack
Reinforcement learning	Support vector machine [25]	Spoofing attack

References

1. Souza E, Wong HC et al (2013) End-to-end authentication in under-water sensor networks. In: 2013 IEEE symposium on computers and communications (ISCC). IEEE, pp 299–304
2. Hu YC, Perrig A, Johnson DB (2003) Packet leashes: a defense against wormhole attacks in wireless networks. In: Proceedings of the 22nd annual joint conference of the ieee computer and communications societies (INFOCOM 2003), April 2003, pp 1976–1986
3. Karlof C, Wagner D (2003) Secure routing in wireless sensor networks: attacks and countermeasures. Ad Hoc Netw 1(2–3):293–315
4. Yuan C et al (2015) A low computational complexity authentication scheme in underwater wireless sensor network In: 11th IEEE international conference on mobile ad hoc and sensor networks, MSN 2015, pp 116–123
5. Kong J et al (2005) Low-cost attacks against packet delivery, localization and time synchronization services in under-water sensor networks. In: Proceedings of the 2005 ACM workshop on wireless security, Cologne, Germany, pp 87–96
6. Ma B (2009) Cross-layer trust model and algorithm of node selection in wireless sensor networks. In: In: International Conference on Communication Software and Networks, ICCSN'09. IEEE, pp 812–815
7. Dini G, Lo Duca A (2011) A cryptographic suite for underwater cooperative applications. In: Proceedings of IEEE international symposium on computers and communications, Kerkyra, Greece, June 2011, pp 870–875
8. Ateniese G et al (2015) SecFUN: security framework for underwater acoustic sensor networks. In: Proceedings of the MTS/IEEE OCEANS 2015—Genova: discovering sustainable ocean energy for a new world
9. Yun CW et al (2016) Ticket-based authentication protocol for underwater wireless sensor network. In: International conference on ubiquitous future networks, ICUFN, Aug 2016, pp 215–217
10. Han G et al (2015) Secure communication for underwater acoustic sensor networks. IEEE Commun Mag 53(8):54–60
11. York N et al. Guide to elliptic curve cryptography. Springer
12. Zhang F et al (2004) An efficient signature scheme from bilinear pairings and its applications. In: Lecture Notes Computer Science, including subseries Lecture Notes in Artificial Intelligence (LNAI) and Lecture Notes in Bioinformatics (LNBI, vol 2947, pp 277–290

13. Boneh D, Shacham B, Boneh H (2004) Short signatures from the weil pairing. J Cryptol 17 (4):297–319
14. Wan C et al (2018) Non-interactive identity-based underwater data transmission with anonymity and zero knowledge. IEEE Trans Veh Technol 67(2):1726–1739
15. Liu CG et al (2012) Iterative key distribution based on MAD neighborhood in underwater mobile sensor networks. Comput J 55(12):1467–1485
16. Luo Y et al (2016) RSS-based secret key generation in underwater acoustic networks: advantages, challenges, and performance improvements. IEEE Commun Mag 54(2):32–38
17. Lal C, Petroccia R et al (2016) Secure underwater acoustic networks: current and future research directions. In: 3rd underwater communication network conference Ucomms, 2016
18. Kulhandjian H et al (2014) Securing underwater acoustic communications through analog network coding
19. Ahmed MR et al (2015) A novel algorithm for malicious attack detection in UWSN. In: 2nd international conference electronic engineering information communication technology iCEEiCT 2015, May 2015
20. Shu T et al (2014) Multi-lateral privacy-preserving localization in pervasive environments. In: IEEE conference on computer communications, IEEE INFOCOM 2014, pp 2319–2327
21. Sastry N et al (2003) Secure verification of location claims. Proc Work Wireless Secure (Section 2):1–10
22. Liu D et al (2005) Attack-resistant location estimation in sensor networks. In: 4th international symposium on information processing sensor networks, IPSN, pp 99–106
23. Zhang R et al (2010) Wormhole-resilient secure neighbor discovery in underwater acoustic networks. In: Proceedings of the IEEE INFOCOM 2010
24. Dini G et al (2011) SeFLOOD: a secure network discovery protocol for underwater acoustic networks. In: Proceedings of the IEEE symposium computational communication 2011, pp 636–638
25. Zhang X, Lu Z, Kang C (2003) Underwater acoustic targets classification using support vector machine. In: Proceedings of the 2003 international conference neural networks signal processing, ICNNSP'03, pp 932–935

Chapter 5
Applications of UWSN

The growing development and research interest in underwater wireless sensor network have led to the implementation of various underwater applications. The usefulness of applications developed underwater is ranging from underwater surveillance to the development of underwater city. The "Challenges of UWSN" really makes it difficult to architect and deploy a highly secured and energy-efficient UWSN. But the cost of UWSN production will be less if compared with the vast benefits provided by underwater applications. The water is the biggest resource available to mankind. To make use of these least used resource, its required to technologically explore seabed's for mankind and aquatic life benefits like disaster monitoring, saving coral-life, industrial and Internet pipeline monitoring and so on. Till date lots of research work and development are done on underwater applications for different domains like surveillance, environmental monitoring, deep sea mining and undergoing prototype projects like "Microsoft Natick" and "Underwater City". There is a list of applications developed using UWSN among which few cream applications are discussed in below subsections.

5.1 Surveillance Applications

The underwater surveillance applications are a boon to military and naval forces. This application helps surveillance of water territory, detection of enemy submarine and mine. There are various underwater sensor networks which are deployed real time in countries like USA and Europe.

P. N. Mahalle et al., *The Underwater World for Digital Data Transmission*,
SpringerBriefs in Computational Intelligence,
https://doi.org/10.1007/978-981-16-1307-4_5

5.1.1 Survelling Water Territory of Country via UWSN

The UWSN plays an important role for naval applications. The building blocks of surveillance network are different types of sensors, AUNs, ROVs, submarines and an offshore station. The different sensors mean specially designed military sensors with cameras, metal detectors and sonars. There are mainly two aims for developing UWSN for surveillance: the first is to preprevent the disastrous actions launched by enemy's and second to develop economical solutions for naval.

The basic application developed was surveillance of underwater territory of nations via UWSN to detect enemy submarines or autonomous vehicles and hidden mines. In mine detection, the mines are made of metallic surface which can be detected by a metal detecting sensor placed on nose of AUV. These detected data or high-resolution image of seafloor with mines can be transmitted to an onshore station by UWSN or through the AUV itself. The UWSN is also used for detection of enemy submarine or battleships for near shore surveillance. In [1], the researchers have designed a different architecture layout where sensor is deployed in a way to cover maximum surveilling area, and then further, data mining techniques are used to classify the object whether submarine, ship, mine and vehicles. In generalize, acoustic communication medium is most commonly used due to its property of longest distance coverage. The combination of radio and acoustic communication medium is also preferred to facilitate faster data transmission. The 1D architecture is used for stationary located mines, whereas 2D and 4D architectures are used for submarine detection. The machine learning approach is also used to classify the mine likes object (MLOs) by using synthetic and semisynthetic data sets for training and testing [2]. In Chesapeake Bay, the network with five AUVs and two sonar types was successfully experimented to detect different types of mines [3]. There are also various other techniques developed to analyse the underwater data such as pattern analyser [4] to detect mines or similar structure objects by using a large mine images database provided by "Naval Surface Warfare Centre".

5.1.2 Submarine Target Localization

Another economical solution by using UWSN for surveillance is submarine target localization. The experimentation in Italy named GLINT09 [5] concludes that single or network of AUVs can be used to localize submarines with lesser risk and manpower. Then for locations where submarines were not detected by sonar systems the researchers in [6] proposed a distributed UWSN with low complexity sensors for submarine detection. The detection data was in the form of binary which represented the presence or absence of submarine. Another approach for target localization was by receivers estimating time-of-arrival (TOA), direction-of-arrival (DOA) and time-difference-of-arrival (TDOA). The receivers with the help of DOA can determine a straight line on which the target is located. The TOA and TDOA are used by receivers to predict that target is on eclipse with source and receiver as foci. The only one source and receiver pair can use information of DOA, TOA and TDOA for target localization [7, 8]. There is a concept of low-visibility targets, which means a single submarine hides itself from sonar systems and becomes invisible. The detection and localization of such low-visibility targets can be done by using UWSN with densely deployed sensors which are cheap in cost.

5.1.3 European Project UAN

There is increase in trend of developing number of infrastructures near and on sea like wave-based power plants or oil and gas drills. These complex structures are situated very far from coast, and they have some infrastructure parts floating in water and some above-water. Thus, such delicate infrastructures are required to be protected from intrusion or natural disasters like tsunami and hurricanes. Therefore, in such scenarios, only deploying UWSN with security algorithms is not sufficient. Rather, it requires an integrated approach of aerial and underwater defence networks. In such applications, the network flexibility, rapid response, faster deployment, data security and automation play a key role. The UAN11 sea trial [9, 10] was the first which practically implemented security in underwater networks and showed the importance of underwater data security. The project implemented network security at middleware supporting both radio and acoustic communications, henceforth increasing the underwater network modularity and flexibility. The European project underwater acoustic network [EU-UAN] was a funded project developed with an aim to protect offshore and coastline infrastructure. The UAN experiment was ended in 2011 at Norway with last sea trial named as UAN11. They used two remotely operated AUVs for surveillance of four fixed nodes which were deployed on site and a mobile node. The UAN11 was an innovative concept for combing underwater and above-water sensors (aerial and surface sensors) in a unique communicative way, i.e. global positioning system [10]. The network was proven to be robust to modifications in topology like addition or deletion of any node. The network performance was calculated using packet-loss, round-trip time and average delivery ratio. The network was functional for five days from 23rd May to 27th May, with continuous node addition and subtraction from network. The acoustic channel variation and addition of security feature greatly and minorly degraded the performance of network. The noting point is that the five fixed nodes and one mobile node were recovered back for recharging their batteries and then redeployed.

5.1.4   Seaweb

The rice Seaweb [11] is said as organized underwater wireless network providing "C3N". The "C3" stands for command, control, communications and "N" is for navigation of autonomous systems. The Seaweb networks are implemented using digital signal processing DSP-based telesonar acoustic modems to interconnect mobile and fixed sensors. The three pillars of such networks are "Seaweb backbone", which includes AUVs, fixed surveillance nodes and repeaters. Then second is "Seaweb Peripherals" which includes unmanned underwater vehicles (UUVs), crawlers, gliders and sonar projectors. The third pillar is "Seaweb Gateways", which connects to command centres that can be floating, submerged or on shore. The gateway also helps to interface Seaweb to terrestrial (TWSN), airborne (sensors in patrol aircrafts) and space-based (satellite) networks. The best example for gateways is surface buoys which are a radio-acoustic communication interface that permits satellites and aircrafts to access the submerged UWSN. The Seaweb networks are used in situations where cabled array of sensors or the array of buoys is vulnerable to network attacks (pilfering) or ship traffic. The Seaweb can also be used in underwater applications requiring wide area coverage making large oceanographic nodes grid and nodes are with limited battery power. The functionality provided by Seaweb networks is localization, navigation and ranging, and hence, it supports the mobile nodes like AUVs, submarines and UUVs. The Seaweb operations are used for UWSN development by U.S. Navy for developing surveillance applications.

5.2   Environment Monitoring

The monitoring applications consist of densely deployed underwater sensors used to monitor the surrounding physical entities or to note any sudden change in underwater environment. The monitoring applications are directly related to life of aquatic world and human beings. The underwater life-saving monitoring applications are classified into (i) water quality monitoring system, (ii) tsunami monitoring system, (iii) aquatic monitoring system.

5.2.1   Water Quality Monitoring Systems

The portable water is a precious resource for all leaving things present above and below the water surface. Therefore, monitoring water quality is an important aspect. The UWSN has proven to be boon for water quality monitoring applications deployed from small lakes to oceans. The UWSN used for water monitoring use special underwater sensors which can monitor various different parameters, such as salinity, nitrogen and fluoride level, pH, temperature and so on. The cost of sensors for nitrogen and fluoride monitoring is very high as compared to pH and temperature monitoring. The UWSN can detect the contaminated water areas so as to prevent the contamination of complete water areas. For example, the Dubai Municipal Corporation has deployed UWSN with its offshore station present at Dubai corporation building. The aim of continuous water monitoring was to save the rare species of birds which drink water from that coastline in Dubai. The fact is that detection of water contaminated zone can save life of various aquatic species and also humans life from waterborne diseases.

The water quality monitoring applications are used in many domains, like in aquaculture to save trout farms [8] where managing the quality of water is of paramount importance to save the fishes in pool. In [12], the authors have developed a wireless network for continuous water quality monitoring of rivers in India. The nodes are designed to mainly monitor the pH of water, which is an important factor affecting water quality. The Zigbee communication is used for wireless transmission of sensed pH value. The sensors used in project are cost-effective as well as energy-efficient. There are joint water quality monitoring research projects going on by Government of India and UK, known as "India-UK Water Quality Programme" [13]. These projects incudes, (i) manufacturing of low-cost optical sensors for monitoring the water quality, (ii) development and implementation of sensors and treatment technology for freshwater in India, (iii) fate and management of emerging contaminants and many more such projects. Therefore, monitoring water quality with high-grade networking technologies is the need of era with increasing pollution.

5.2.2 Tsunami Monitoring Systems

The underwater natural disasters like volcanoes, earthquakes and tsunamis can endanger the aquatic life, humans and also underwater oil and gas industries. These natural disasters are not predictable, and hence, its required to continuously monitor the geological changes occurring below water surface. The UWSN is only way to monitor such conditions and to give alarming signals accordingly. The noting point is the UWSN architectures developed for monitoring will also include a satellite for faster broadcasting of sensed data to various base stations. In traditional approach, the scientist community relied on surface buoys to predict the path of seismic waves. The UWSN for real-time tsunami monitoring systems consists of sensor nodes spread over the entire seabed. The warning signal is generated when a large volume of wave water moves in short period of time, i.e. increased in sea level. These systems generally follow 3D architectures which are combination of UWSN, satellite network and TWSN. The sensor nodes are organized in clusters, and then these cluster members communicate to neighbouring nodes over short distances, whereas the cluster heads transmit collected data to the nearby buoy. Then the surface station is responsible to collect data from all cluster heads and further transmit it to costal data collection centre or satellite network. The collection data centres can also send data to global centres through satellites. The collected data at base stations is then analysed by an analyst or using a dedicated software which generates warning alarms in accordance with collected data.

In state of the art, an 4D-UWSN architecture is been proposed for early warning signal generation [14] in case of tsunamis. The OFDM modulation technique and multi-carrier communication are used for efficient underwater communication. In [15], the authors have proposed an UWSN architecture using "seismic pressure sensors", which predicts seismic waves and also relay data to surface stations using routing protocols. They have divides the network using three node types, named as sensor node, barrier and commander. The UWSN consists of eighty seismic pressure sensors which sense pressure data and then relay it to specially designed nodes known as commander. These nodes are trained to analyse and predict the data and then if required the four barriers are needed to be fired. The barriers are used to reduce the effect of wave but they cannot stop a tsunami. In [16], there is a survey on different underwater disaster monitoring applications. In real time, there is lots of theoretical research done on UWSN applications but very few deals with a complete design addressing all issues like reliability, robustness, security and energy efficiency.

5.2.3 Aquatic Monitoring Systems

The use of UWSN is also done to develop aquatic monitoring applications with an aim to reduce the life risk of divers and to save rare aquatic species and coral reef. The deployment of sensors in such applications is done in distributed manner and with a limited transmitting range. The multi-hop data communication and robust network topology are the key features of underwater networks for aquatic monitoring. An European project is designed for underwater aquatic monitoring. The project aimed to monitor aquatic environment like path of fishes and their activities and for geological surveys. The protocol used for data transmission among sensor node was TDMA-based. In [17], there is a framework developed using UWSN used for aquatic monitoring. The framework is designed to be scalable, robust to topological changes and embedded energy-aware mechanisms in gateway nodes and sensor nodes. The main aspect of any aquatic monitoring applications is data storage and transmission, sensing, data visualization and firing warning signals whenever required. The developed framework leads to following outputs like, (i) ability to simultaneously monitor the aquatic environment at multiple levels with an intention to verify biological models, (ii) to prediction of growth of toxic algal, (iii) predict deaths of aquatic species due to climatic changes like cyclone formation, (iv) availability of precious time-critical data for triggering alarms and in response managing the marine resources. This monitoring system is deployed in Queensland, Australia, for monitoring water temperature and luminosity with a vision to monitor coral-life. The topology of distributed UWSN is star and is based on clusters, where the cluster member transmits data to gateway nodes which further transmit to control centres. The framework guarantees quality-of-service, energy efficiency and optimal use of solar energy. The authors in [18] focus on development of inexpensive aquatic monitoring application based on smart environmental monitoring and analysis technology. This technology is said as smart WSN which can select most appropriate technology for aquatic monitoring and marine research. The UWSN used consists of large number of cheaper and plug-and-play intelligent sensors covering a large geographical area and can therefore collect huge amount of marine data. The SEMAT aims to deploy UWSN in shallow water with minimal expertise and to implement short-distance transmissions. Another UWSN, for aquatic monitoring in coastal shallow water was proposed by authors in [19]. The project was deployed in Mar Menor coastal lagoon in Spain. The deployed UWSN consists of numbers of sensors which are responsible to collect oceanographic data and then relay them to sink node mainly buoys. The end station for collected data is a remote station. The authors have proven that deployed network is efficient in terms of data transmission, data reliability and network lifetime. In study done in [20], the marine vehicles (AUV, ROV, UAV) are used as mobile nodes of UWSN with an aim for tracking ocean features and sampling of ocean area. This network was deployed in South China Sea, and their work had also designed protocols for ocean monitoring applications. A detailed study and discussion on use of UWSN technology for aquatic monitoring are provided by authors in [21]. It includes recent advancements and study of different technologies used in aquatic monitoring applications.

5.2.4 Oil Spill Leakage Detection

The oil spills are death taking water pollution occurring due to high interference of humans for oil field discovery and its transportation oil through oceans. The minute leaks in oil pipelines can endanger thousands of aquatic species, and therefore, using underwater wireless network for detection of oil spills is an important application of underwater wireless networks. The early detection of oil spills by using UWSN will further faster the cleaning process indirectly saving the marine life. In [22], authors have proposed a decentralize ad hoc UWSN for the detection of pollution. They have developed a new protocol stack with three new planes, which are used for power management, coordination and localization. The power management plane is used for selecting the most suitable synchronization protocol, whereas the coordination plane is worked for time synchronization among all nodes. The third plane is localization and is responsible for the selection of best neighbour. The main attention is given to synchronization and routing algorithms with an aim to increase network availability and quality-of-service. There is also research done on designing of special sensors which can transmit the information about location and thickness of oil spills [23]. To find the thickness of oil spill, authors have developed two array algorithms named as light sensor and conductivity array. In addition, they have also designed a simulator which shows mapping of sensed data about thickness of oil spill and its location on map.

The oil and its products are main source of revenue for many countries especially for Nigeria. The curse to commercial production of oil is vandalism of pipelines carrying oil. Nigerian economy had a loss of seven billion dollars due to such oil thefts. Other than economy fall, the pipeline vandalism also results in oil spillages into underwater environment. The post-effect of oil spillage is increasing in fish mortality rate and water pollution. Therefore, to handle the issue of worst oil spillage due to pipeline vandalism in Nigeria, the authors in [24] have developed an underwater wireless sensor network to monitor the oil pipelines and marine environment. The UWSN can detect the damaged part of oil pipeline and then will accordingly respond to base stations. The authors have concluded saying UWSN as best choice for detecting oil spills as compared to other networking techniques.

5.3   Microsoft Project Natick

The Microsoft Company developed a prototype project named "Natick" [25]. The project aimed to reduce the cooling cost of data centres by using water as natural cooling agent. The Microsoft noticed and stated that if more than half of world population resides near coasts then, "why not their data resides in water"? The innovative idea of keeping data centres into ocean at natural cooling temperature of 60 °F to save huge electricity bills have somewhere given a push to the development of UWSN with high transmission speed. These successful prototype project revealing that, a day is not far when there will be number of underwater data centres connected wirelessly with the help of UWSN. The world is changing rapidly with technology, and hence, UWSN is future of networking technology.

The data centre is actually a white colour cylinder residing on seabed and containing computers, hard-disk racks. The project was successfully deployed for five years and includes 12 racks consisting of 864 servers with storage capacity of 27.6 petabytes which can sufficiently store five million of movies. Then computing capacity of data centre is equivalent to thousands of desktop machines. The project is implemented in Orkney, and hence, the data centre was supplied with power from by undersea cable and renewable energy from Orkney Islands. The Orkney Islands was chosen because its main centre of renewable energy. In these prototype models' cable is also used to connect underwater data centres with outside Internet. In future, this cabled network can be made wireless using UWSN. To avoid corrosion problem of underwater data centre, a provision is made to absorb all oxygen and water vapour from atmosphere, as no human stay into deep oceans. The only drawback found of Natick was that the damaged computer residing in cylinder is not reparable, but the ray of hope is that the failure rate will be lower as compared to data centres on land.

5.4 Deep Sea Mining

The removal of mineral deposits like aluminium, copper, zinc, magnesium, nickel, cobalt and lithium from the area 200 m below the seabed is known as deep sea mining. According to facts and figures available, the seabed measuring around the size of Mongolia is set aside for sea mining in Indian and Pacific Oceans. In recent years, the data mining field has received greater attention due to high degradation of mineral deposits on land. The research is going on to develop a high-tech technological solution for deep-sea mining. In regard to it, UWSN with combination of mining tools can serve as good solution for deep-sea mining. But such research of using UWSN for deep-sea mining is still in its infancy.

The exploration of mine can also lead to exploitation of sea, and in such situation's, government agencies can think of UWSN to monitor the seabeds. The researchers have developed solution for deep-sea mining by using UWSN and remotely operated vehicles (ROVs) with wireless communication system for fascinating vision base monitoring. This research was mainly done with an aim to find and measure magnesium crust in deep oceans. The researchers have also developed UWSN with a combination of underwater mobile network (i.e. network of AUVs and ROVs) and static UAN for digging and hunting deep-sea mines.

References

1. Cayirci E et al (20016) Wireless sensor networks for underwater surveillance systems. Ad Hoc Netw 4(4):431–446
2. Barngrover C et al (2015) Semisynthetic versus real-world sonar training data for the classification of mine-like objects. IEEE J Oceanic Eng 40(1):48–56
3. Khaledi S et al (2014) Design of an underwater mine detection system. In: IEEE systems and information engineering design symposium (SIEDS), pp 78–83
4. Rao C et al (2009) Underwater mine detection using symbolic pattern analysis of sidescan sonar images. In: Proceedings of the American control conference, pp 5416–5421
5. Hamilton MJ et al (2010) Antisubmarine warfare applications for autonomous underwater vehicles: the GLINT09 sea trial results. J Robot 27(6):890–902
6. Zhou S, Willett P (2006) Submarine location estimation via a network of detection-only sensors. In: Proceedings of the IEEE conference on information science and systems CISS 2006, vol 55(6), pp 363–368
7. Felemban E (2013) Advanced border intrusion detection and surveillance using wireless sensor network technology. Int J Commun Netw Syst Sci 06(05):251–259
8. Yalcuk A (2015) Evaluation of pool water quality of trout farms by fuzzy logic: monitoring of pool water quality for trout farms. Int J Environ Sci Technol 12(5):1503–1514
9. Mooney JG, Johnson EN (2014) A comparison of automatic nap-of-the-earth guidance strategies for helicopters. J Robot 33(1):1–17. Available from: http://onlinelibrary.wiley.com/doi/10.1002/rob.21514/abstract
10. Cresta M et al (2010) Archimede: integrated network-centric harbour protection system. In: International waterside security conference 2010
11. Rice J et al (2001) Seaweb underwater acoustic network. Spaware Navy:234–250. Available from: http://www.spawar.navy.mil/sti/publications/pubs/td/3117/234.pdf

12. Šaliga J et al (2015) Wireless sensor network for river water quality monitoring. In: XXI IMEKO world congress 'Measurement Research Industry', pp 1–7
13. https://iukwc.org/india-uk-water-quality-programme-1
14. Kumar P et al (2012) Underwater acoustic sensor network for early warning generation. In: Ocean 2012 MTS/IEEE harnessing power ocean, June 2012
15. Casey K et al (2008) A sensor network architecture for Tsunami detection and response. Int J Distrib Sens Netw 4(1):28–43
16. Lloret J (2013) Underwater sensor nodes and networks. Sensors (Switzerland) 13(9):11782–11796
17. Alippi C et al (2011) A robust, adaptive, solar-powered WSN framework for aquatic environmental monitoring. IEEE Sens J 11(1):45–55
18. Trevathan J et al (2012) SEMAT—the next generation of inexpensive marine environmental monitoring and measurement systems. Sensors (Switzerland) 12:9711–9748
19. Pérez CA et al (2011) A system for monitoring marine environments based on wireless sensor networks. In: Ocean 2011, IEEE, Spain, July 2011
20. Zhang S et al (2012) Marine vehicle sensor network architecture and protocol designs for ocean observation. Sensors 12(1):373–390
21. Xu G, Shen W, Wang X (2014) Applications of wireless sensor networks in marine environment monitoring: a survey. Sensors (Switzerland) 14(9):16932–16954
22. Khan A, Jenkins L (2008) Undersea wireless sensor network for ocean pollution prevention: a novel paradigm for truly ubiquitous underwater systems. In: 3rd IEEE/create-net international conference communication system software middleware, COMSWARE, Feb 2008, pp 2–8
23. Koulakezian A et al (2008) Wireless sensor node for real-time thickness measurement and localization of oil spills. In: IEEE/ASME international conference advanced intelligent mechatronics, AIM, Aug 2008, pp 631–636
24. Henry NF, Henry ON (2015) Wireless sensor networks based pipeline vandalisation and oil spillage monitoring and detection: main benefits for Nigeria oil and gas sectors. SIJ Trans Comput Sci Eng Appl 3(1):1–7
25. https://natick.azurewebsites.net/

Chapter 6
Conclusion

6.1 Summary

In this book, we have tried to brief on all topics, relating to use of "Underwater World for the purpose of Digital Data Transmission". The book chapters are describing a short story on UWSN showcasing from "What is UWSN?" to "Applications of UWSN". The first introductory chapter speaks about different architecture models in UWSN (1D, 2D, 3D and 4D-UWSN), the components of UWSN and shows the difference between TWSN and UWSN.

The second chapter discuses carriers of digital data in underwater world, i.e. the three "Communication Mediums" (acoustic, radio and optical channel). Also, the comparative study done on communication medium will help researchers to decide the best suitable communication medium for their underwater research projects. The challenges of communication mediums help to understand the stumbling blocks in development of UWSN. The third chapter gives a short description on the "Protocol Stack" of UWSN, defining each layer of underwater network.

The fourth chapter is about security of digital data transferred through the unpredictable underwater world. In state of the art, there are many "Threats and Attacks in UWSN", which we have further classified according to protocol layers, capability of attacker and attacks done when data is in transit phase. These chapter also discuss cross-layer approach for confirming data security and a short summarization on various different attack resistant strategies.

The fifth chapter briefs on application of UWSN in various domains like surveillance, environmental monitoring and the projects in their prototype phase. And at last, this chapter will give a complete overview of book with a glance on the open research issues and future of UWSN.

© The Author(s), under exclusive license to Springer Nature Singapore Pte Ltd. 2021   65
P. N. Mahalle et al., *The Underwater World for Digital Data Transmission*,
SpringerBriefs in Computational Intelligence,
https://doi.org/10.1007/978-981-16-1307-4_6

6.2 Open Research Issues

Each single research study done in UWSN can be said as incomplete. These is because the field of UWSN has received researcher attention and interest in post-2000 and hence till date there is no standardization or benchmarks in domain of underwater wireless networks. Therefore, it is a long list of open research issues in UWSN out of which we will focus on some.

1. The security of UWSN is the least focused area in spite of vulnerable underwater environment due to the use of acoustic communication medium.
2. The 90% of UWSN which are real-time deployed or proposed theoretically have used acoustic communication medium. The very few research works are done on use of optical and radio waves as primary communication medium of UWSN.
3. The only few researchers countable on fingertips have worked on, cross-layer approach in UWSN to prevent same attacks occurring in two or more protocol layers. Hence, leaving cross-layer approach, an open research issue to smartly tackle the attack and threats in UWSN.
4. The underwater sensor designing, development and deployment is also an open research issue.
5. The architectures developed for UWSN mainly consists only underwater acoustic sensor nodes as its primary component. These underwater architectures can be further enhanced by using permutations and combinations of mobile and static underwater nodes or an architecture with a combination of acoustic and optical sensors or the use of AUV and ROV with static and mobile UWSN.

6.3 Future of UWSN

The underwater world is changing rapidly due to increase is underwater applications and in its technological use. From our study, it can be stated that in future, "Underwater Wireless Sensor Network is going to be as Part of Oceans".

To give a vision on future of underwater wireless networks, we will like to share some upcoming underwater projects like underwater data centres, underwater city, underwater drones, use of IoTs in underwater and the most recent project is on removal of leaking mercury form German World War II submarine wreck.

Underwater Data centres: The availed high-speed data to coastal areas all over world is possible due multiple data centres fitted on seabeds and connected wirelessly to outside internet. Hence, future of UWSN is underwater data centres connected using underwater wireless network.

Underwater City: The Japanese company has declared; that in five years, it will develop an underwater city accommodating five thousand people with a facility of five-star underwater hotels and multiplexes. This project tells about the advancements in underwater technology and the future of underwater world.

The world is changing and so is the technology and we as researcher is witnessing it. The fascinating, dynamic and challenging field of UWSN is attracting the attention of many researchers, and in future, it is also going to be a routine research network like TWSN and IoTs.

Printed in the United States
by Baker & Taylor Publisher Services